AF866361

The End of Cosmic Anomalies

Sandeep Chavan

Published by Gyrus Vision, 2026.

THE END OF COSMIC ANOMALIES

First edition. June 1, 2026.

ISBN: 978-8168977686

Written by Sandeep Chavan.

Table of Contents

To the restless intelligence that refuses to stop at inherited answers.

To every observer who has looked at the sky and felt that the universe is deeper than the language used to explain it.

And to those who still believe that questioning is not rebellion, but the beginning of honest understanding.

"The universe does not become anomalous when it refuses our model.

The model becomes incomplete when it refuses the universe."

— *Sandeep Chavan*

Opening Note

Why This Book Had to Be Written

There are moments when science does not fail, but its language becomes too small.

The universe does not panic when our equations tremble. Galaxies do not become anomalies because our timelines cannot accommodate them. Black holes do not become mysterious because our imagination stops at singularity. The cosmos does not break merely because our models begin to crack.

And yet, whenever observation refuses to obey our expectation, we quickly call it a crisis.

The James Webb Space Telescope did not create this crisis. It only revealed it.

For decades, modern cosmology has been presented as if the universe is a long historical movie reel. We were told that when we look deeper into space, we are looking further back into time. We were told that the early universe must be small, immature, chaotic, and structurally incomplete. We were told that galaxies require long mechanical growth, that black holes must grow through accumulation, and that the universe is expanding like a fabric stretched by invisible forces.

Then JWST opened its eyes.

Suddenly, the so-called early universe appeared too mature. Galaxies looked too bright, too structured, too massive, too soon. Supermassive black holes appeared where they were not supposed to appear. Redshifted regions refused to behave like simple historical fossils. The Hubble tension remained alive. The cosmic timeline, which had appeared elegant from a distance, began showing stress marks under sharper observation.

Mainstream science responded in its usual way.

It did not immediately ask whether the foundation itself needed reconsideration. Instead, it began adding explanatory patches. More efficient star formation. Exotic early stars. Heavy black hole seeds. Modified dark energy. Adjusted parameters. New possibilities placed carefully inside old assumptions.

This is not wrong as a working method. Science must attempt to preserve a model until it becomes impossible to preserve it honestly.

But there comes a point where patches no longer protect the model. They expose it.

This book begins at that point.

I am not writing this book to attack science. I am writing it because science deserves better than decorative repairs around an exhausted interpretation. Science is not loyalty to inherited language. Science is loyalty to reality, even when reality refuses to fit inside familiar mathematics, famous names, or institutional comfort.

The problem is not observation.

The problem is interpretation.

What if the universe is not presenting anomalies?

What if our interpretation has become too chronological, too mechanical, too dependent on the idea that cosmic depth must mean cosmic past?

What if JWST is not looking at dead history but penetrating deeper layers of the present universe?

What if redshift is not only a sign of distance and expansion, but also a sign of field strain, ripple tension, and structural resistance?

What if galaxies do not grow only through slow particle aggregation but emerge through field realignment inside Structured Vacuum Energy?

What if black holes are not infinite-density monsters but saturation zones where ripple coherence collapses beyond observational recovery?

What if time itself is not a road on which the universe travels, but a measurable delay created by the alignment of energy states?

These questions are not casual imagination for me. They arise from the broader framework I call Universal Energy Dynamics, or UED.

UED does not begin with particles as the final truth. It does not begin with empty spacetime waiting to be bent. It does not begin with time as an independent cosmic highway. It begins with a deeper proposition: existence is a continuous consequence of universal intelligence expressing itself through structured energy, equilibrium-seeking behavior, ripple interaction, and field resolution.

In this framework, matter is not a dead object. It is a field consequence.

Gravity is not a mysterious pull from nowhere. It is an emergent outcome of energy gradients seeking balance.

Time is not an invisible river flowing through the universe. It is Alignment Threshold Time, the delay required for energy states to resolve into measurable consequence.

A black hole is not a physical hole in reality. It is a Ripple Saturation Zone, a region where field coherence collapses from the standpoint of external observation.

And the universe is not a broken machine producing anomalies. It is a structured, intelligent, equilibrium-seeking whole that appears strange only when we insist on reading it through the wrong lens.

This book is therefore not merely about JWST.

JWST is the trigger.

The real subject is interpretation.

For centuries, humanity has confused measurement with meaning. We developed powerful instruments, then placed their observations inside stories we had already accepted. When observation matched the story, we called it confirmation. When observation resisted the story, we called it anomaly.

But perhaps an anomaly is not always a problem in the universe.

Sometimes an anomaly is a message from reality saying: your story is too small.

The so-called cosmic anomalies are not random disturbances. They are signals. They are stress points where the older model begins to reveal its hidden assumptions. They show us where we have mistaken mathematical convenience for physical truth, and where we have treated human narration as cosmic fact.

This book does not ask the reader to reject everything science has built. That would be immature. Every model has contributed something. Every serious theory is a step in the long struggle of intelligence to understand existence.

But this book does ask the reader to pause before accepting the word "anomaly."

Anomaly according to whom?

Impossible under which assumption?

Too early compared to what timeline?

Too massive according to which growth story?

Too strange for which model?

The universe is not obligated to follow our intellectual comfort.

If our timeline cannot explain observed structure, the structure is not guilty. If our theory cannot explain maturity at high redshift, the galaxy is not confused. If our model cannot explain central galactic regulators, the cosmos is not violating the law. It is our law that may be incomplete.

This is the spirit in which this book must be read.

Not as rebellion for the sake of rebellion.

Not as rejection for the sake of attention.

Not as an attempt to replace one dogma with another.

This book is an invitation to look again.

To look at galaxies not merely as ancient objects, but as field consequences.

To look at black holes not as cosmic endings, but as regulatory thresholds.

To look at redshift not only as distance, but as resistance.

To look at time not as the ruler of reality, but as the mind's way of arranging consequences.

To look at the cosmos not as a movie playing from beginning to end, but as a structured depth of the continuous Now.

The end of cosmic anomalies does not mean the end of mystery.

Mystery will remain. It must remain. A universe without mystery would be a dead textbook, not a living reality.

But there is a difference between mystery and confusion.

Mystery invites deeper intelligence.

Confusion arises when we force reality into the wrong frame.

This book is written to remove that confusion.

If Universal Energy Dynamics is correct even as a serious interpretive framework, then many so-called cosmic anomalies are not anomalies at all. They are natural expressions of energy redistribution, ripple saturation, structural equilibrium, and alignment delay.

The cosmos was never broken.

Perhaps we were only watching it through a cracked mirror.

And now, with JWST forcing us to confront what our older assumptions cannot comfortably explain, the mirror has begun to crack further.

That is not the end of science.

That may be the beginning of a more honest one.

Preface

Every age believes it has finally understood the universe.

Then the universe quietly proves that understanding is still incomplete.

This book begins from that humility. It does not begin with the claim that modern science is useless. It does not begin with rejection. It does not begin with anger against established cosmology. It begins with a simple observation: the sharper our instruments become, the more visible our assumptions become.

The James Webb Space Telescope has given humanity one of the most powerful moments in modern cosmic observation. It has looked deeper, resolved fainter signals, and revealed structures that have forced scientists, thinkers, and observers to reconsider what they expected from the so-called early universe. Galaxies appear where they were expected to be less mature. Compact red structures disturb simple formation stories. Black-hole-like regions challenge slow growth assumptions. Redshift continues to carry enormous interpretive weight. Hubble tension remains unresolved enough to keep asking deeper questions.

The usual response is to call these things anomalies.

But this book asks whether the word "anomaly" itself needs to be examined.

An anomaly is not merely something strange. It is something that does not fit the frame through which it is interpreted. If the frame is too narrow, reality will naturally appear uncomfortable. If the language is too old, new observation will sound like contradiction. If the model carries hidden assumptions, sharper data will expose them.

This book is written from that point of exposure.

The central argument is simple: the universe may not be broken. Our interpretation may be incomplete.

Universal Energy Dynamics, or UED, provides the framework through which this book reopens the meaning of matter, gravity, black holes, redshift, time, cosmic depth, and Hubble tension. UED does not begin with dead matter floating inside empty space. It does not begin with time as an independent cosmic road. It does not begin with black holes as monsters or redshift as a single-purpose clock.

It begins with structured energy, field condition, consequence, resistance, alignment, saturation, and equilibrium-seeking behavior.

From this starting point, the universe appears differently.

- Matter becomes ripple condensation within Structured Vacuum Energy.
- Gravity becomes energy-gradient equilibrium.
- Black holes become Ripple Saturation Zones.
- Galactic centers become central regulators.
- Redshift becomes field strain and resistance.
- Time becomes Alignment Threshold Time and conscious narration.
- Cosmic depth becomes field depth, not merely pastness.
- Hubble tension becomes state-dependent resolution delay.

These ideas are bold, but they are not offered as slogans. They are offered as an interpretive lens. The purpose of this book is not to close the discussion, but to open it in a deeper direction. A framework must earn its strength through clarity, criticism, refinement, and its ability to make sense of observations without forcing them into artificial discomfort.

Readers should therefore approach this book with two qualities at once: openness and discipline.

Openness is needed because no major shift in understanding begins inside total comfort. If every idea is rejected simply because it does not use familiar language, then knowledge becomes a museum of old confidence.

Discipline is needed because every new idea must be examined carefully. A new framework should not be accepted merely because it sounds interesting. It must be questioned. It must be tested. It must be compared. It must grow through honest challenge.

This book asks for neither blind acceptance nor blind rejection.

It asks for serious reading.

The book is structured in three movements.

First Part exposes the crisis of interpretation. It shows how modern cosmology has become surrounded by tensions, exceptions, patches, and assumptions, especially after JWST began disturbing the comfort of the cosmic timeline.

Second Part introduces the UED reinterpretation. It reopens matter, gravity, black holes, galactic centers, and redshift through a field-first understanding.

Third Part dismantles the cosmic movie reel. It questions Hubble tension, time, motion, history, depth, and the habit of treating the universe as a film moving from beginning to present.

The final chapter gathers everything together and argues that cosmic anomalies may cease to be anomalies when seen through a deeper lens.

This book is not written only for scientists. It is written for serious readers, thinkers, students, researchers, philosophers, and anyone willing to look at the universe without surrendering intelligence to inherited comfort. Some readers may agree. Some may disagree. Both responses are welcome if they are honest.

The universe does not need followers. It needs readers with courage.

A reader may disagree with UED and still gain something from this book: the habit of separating observation from interpretation. That alone is valuable. Once this separation becomes clear, many intellectual confusions begin to weaken. We stop asking only whether the data fits the model. We also ask whether the model is large enough for the data.

That is the real spirit of this work.

The cosmos has never promised to obey our timelines. It has never promised to behave like a movie reel. It has never promised that depth must mean pastness, that maturity must mean age, or that darkness must mean emptiness. These are our interpretations. Useful, perhaps. Powerful, perhaps. But not untouchable.

The universe remains greater than our language. And when language becomes too small, reality does not shrink.

The language must grow. This book is one attempt to help that growth.

How to Read This Book

This book should not be read as a casual rejection of mainstream cosmology.

It should be read as a serious reinterpretation.

The observations discussed here—JWST discoveries, redshift, galaxies, black-hole-like regions, and Hubble tension—are not dismissed. They are accepted as important signals. What this book questions is the meaning assigned to them.

Read slowly where the argument challenges familiar ideas. The book does not ask you to abandon science. It asks you to separate observation from interpretation.

When you see the word "anomaly," pause.

Ask: anomaly under which model?

When you see "early universe," ask: early according to which timeline?

When you see "black hole," ask: object, or field saturation?

When you see "redshift," ask: distance only, or field strain too?

When you see "time," ask: cosmic road, or alignment delay?

This book is built progressively. It exposes the crisis of interpretation. It introduces the UED lens. And it questions the cosmic movie reel. The final chapter brings the argument together.

You do not need to agree with every conclusion immediately. Agreement is not the first requirement. Serious attention is.

Read with curiosity, not fear.

Read with discipline, not blind acceptance.

Read with the willingness to ask whether the universe is truly strange—or whether the frame through which we read it has become too small.

Part I — The Crisis of Interpretation

"The universe does not become strange when observation improves. Only our assumptions become visible."

Chapter 1 — The Universe Did Not Become Strange Overnight

Why modern cosmology is increasingly surrounded by tensions, exceptions, patches, and anomalies.

When Reality Refuses the Script

The universe did not become strange overnight.

It did not suddenly wake up one morning and decide to violate our equations. It did not begin producing mature galaxies too early, massive structures too soon, or black holes too large merely to embarrass our theories. The cosmos has always been doing what it does. It has always been operating through its own deeper order, its own silent discipline, its own structural intelligence.

What has changed is not the universe.

What has changed is our ability to observe it.

For a long time, humanity looked at the sky through limited instruments and unlimited imagination. We built stories from light that reached us, from patterns we could detect, from mathematics we could trust, and from assumptions we were not always ready to question. Slowly, these stories became models. The models became frameworks. The frameworks became language. And eventually, language became confidence.

This is natural.

Human intelligence cannot work without structure. We need models to think. We need timelines to arrange events. We need categories to make sense of overwhelming reality. Without interpretation, observation is only scattered data. So science created a grand cosmic story: a beginning, expansion, cooling, particles, atoms, stars, galaxies, planets, life, and observers.

It was elegant.

It was powerful.

It was useful.

But usefulness is not final truth.

A map can guide us for a long distance and still become unreliable when the terrain changes. A theory can explain many observations and still fail when observation becomes sharper. A framework can dominate intellectual culture for decades and still carry hidden assumptions that remain unnoticed until reality presses against them.

This is what is happening now.

The James Webb Space Telescope did not break the universe. It only improved the mirror through which we look. And when the mirror became sharper, the old picture began to look uncomfortable.

Suddenly, the so-called early universe did not look as immature as expected. Some galaxies appeared too bright, too structured, too massive, too soon. Some black hole candidates seemed too developed for the available chronological window. The Hubble tension did not disappear like an instrumental error. Redshifted regions did not behave like simple historical leftovers waiting obediently inside the standard timeline.

Then the word "anomaly" entered the room again and again.

But we must be careful here.

An anomaly is not automatically a defect in reality. It is often a defect in interpretation. It is the point where the universe refuses to follow the script we have written for it. It is reality tapping on the wall of our model and saying, "Look again."

Modern cosmology is increasingly surrounded by tensions, exceptions, patches, and crisis language not because the universe has become chaotic, but because our inherited interpretive structure is under pressure. The old model still has value, but its comfort is no longer enough. Observation has moved ahead. Interpretation must now catch up.

This chapter begins from that simple but powerful realization.

The universe was never obligated to satisfy our timeline.

If a galaxy appears too mature for our model, perhaps the galaxy is not the problem. If a black hole appears too early in our calculation, perhaps the calculation is not reading the event correctly. If distant measurements and local measurements disagree, perhaps the disagreement is not merely an error, but a signal of deeper field behavior.

The cosmos does not need permission from theory to exist as it is.

The problem begins when human intelligence becomes more loyal to the story than to the signal.

And that is why the universe did not become strange overnight.

It only became harder to misread.

1.1 — The Comfortable Universe We Were Told to Imagine

Every age creates a universe it can understand.

Ancient minds imagined the universe through gods, heavens, cycles, omens, and divine order. Later, philosophical minds imagined it through principles, elements, harmony, and purpose. Modern scientific minds imagined it through expansion, particles, forces, spacetime, equations, and measurable timelines.

This was not wrong.

It was the natural growth of human intelligence.

As observation improved, imagination became more disciplined. The telescope changed the sky from a mythological ceiling into a physical depth. Mathematics gave structure to wonder. Physics brought motion, light, gravity, matter, and time into one grand conversation. Slowly, humanity began to feel that the universe could be read like a book.

And from this confidence emerged a very comfortable story.

The universe began.

It expanded.

It cooled.

Particles formed.

Atoms formed.

Stars appeared.

Galaxies slowly gathered.

Planets emerged.

Life eventually became possible.

Consciousness appeared.

And finally, human beings looked back and reconstructed the whole story.

It is a beautiful narrative. It has rhythm. It has sequence. It has intellectual satisfaction. It gives us the feeling that existence has been arranged chapter by chapter, like a cosmic biography written from the first moment to the present day.

This is why the standard cosmological model became so powerful. It did not merely offer equations. It offered a storyline. It gave the mind something it loves deeply: order.

The human mind is uncomfortable with scattered reality. It wants beginning, middle, and end. It wants cause and effect in a straight line. It wants childhood before maturity, simplicity before complexity, darkness before light. So when modern cosmology arranged the universe into an evolutionary timeline, the mind accepted it easily.

In that timeline, deeper space became older time.

Farther light became ancient light.

High redshift became early universe.

Early universe became immature universe.

And immature universe was expected to contain smaller, younger, less organized structures.

This became the script.

Once the script was accepted, every observation was placed inside it. If a galaxy appeared far away, it was treated as old in light but young in cosmic development. If a structure appeared at high redshift, it was expected to be primitive. If a black hole existed in such a region, it had to be explained as something that grew from earlier seeds within a limited chronological window.

The assumption was simple: the universe must have built itself slowly, mechanically, step by step.

This assumption became so natural that it almost disappeared from attention. It became background thinking. Scientists did not always need to state it because the model itself carried it silently. The cosmic timeline was not merely a theory anymore; it became the mental stage on which all interpretation happened.

That is where the danger begins.

A model is useful when it helps us read reality. But it becomes restrictive when reality must first pass through the model before it is allowed to be understood.

The comfortable universe we were told to imagine was not only scientific. It was psychological. It gave us a universe that behaved like a disciplined student: first small, then large; first simple, then complex; first chaotic, then mature. It gave us the satisfaction of progression.

But reality is not obligated to follow our classroom sequence.

The universe may not be growing according to the neat mental ladder we have built. Structure may not always require the long mechanical childhood we imagine. Cosmic depth may not always mean historical infancy. Maturity may not always be the result of time alone.

This is where modern observation begins to disturb the comfort.

Because when better instruments look deeper, they do not always find the obedient universe expected by the script. They find brightness where dimness was expected. Structure where chaos was expected. Massive formation where slow growth was expected. Regulatory centers where infancy was expected.

And then we become surprised.

But perhaps the surprise does not come from the universe.

Perhaps it comes from the comfort of the story we became too ready to believe.

1.2 — When Observations Started Becoming Uncomfortable

A strong scientific model does not become weak because one observation challenges it.

This must be understood clearly.

Science cannot panic every time a new measurement appears uncomfortable. Reality is complex. Instruments have limits. Data requires calibration. Interpretation requires patience. Sometimes an anomaly disappears after better measurement. Sometimes a contradiction becomes a doorway to refinement. Sometimes a strange observation teaches us that the model was mostly correct but incomplete in one corner.

So the problem is not that modern cosmology has anomalies.

The problem is that the anomalies are no longer behaving like isolated accidents.

They are gathering.

They are repeating.

They are appearing from different directions, through different instruments, in different forms, and around the same deeper assumptions. One tension may be ignored. Two may be adjusted. Three may be explained with technical sophistication. But when many tensions begin circling the same foundation, intelligence must stop and ask a harder question.

Are we facing scattered exceptions?

Or are we seeing the outline of a deeper misreading?

For many years, the standard cosmological framework carried its confidence through a grand chronological expectation. The early universe was expected to be young not only in time, but also in structure. In this view, cosmic maturity required duration. Galaxies had to assemble gradually. Stars had to form, explode, enrich the environment, and create conditions for later complexity. Black holes had to grow from smaller seeds through accretion and mergers. The universe had to earn its structure slowly.

Then sharper observation began to disturb this expectation.

High-redshift galaxies appeared brighter than expected. Some appeared more massive than the timeline comfortably allowed. Some seemed too compact, too organized, or too mature for the age assigned to them. Instead of seeing only weak, primitive, irregular beginnings, observation began showing structures that looked as if the universe had reached organizational capacity earlier than expected.

Then came the difficulty of supermassive black holes.

If a black hole is imagined as an object that grows through the gradual consumption of matter, then a very massive black hole at extreme distance becomes a chronological headache. There seems to be too little time for it to become what it appears to be. The model must then search for heavy seeds, unusually fast accretion, special formation pathways, or rare conditions that allow the object to bypass normal growth.

Again, this may be scientifically interesting.

But the pattern matters.

Where the timeline struggles, extra mechanisms are introduced.

The same discomfort appears in the Hubble tension. Measurements of the expansion rate from the nearby universe do not comfortably match those inferred from the early universe under the standard framework. This is not a small poetic disagreement. It touches the basic reading of cosmic scale, age, expansion, and the assumed uniformity of the model. For years, it was hoped that the tension might dissolve as a measurement issue. But it has remained stubborn enough to be treated seriously.

Then there is the silent dependency on dark matter and dark energy.

Modern cosmology does not directly observe these in the ordinary sense in which we observe familiar matter. They are inferred because the model requires them to explain motion, structure, gravitational behavior, and cosmic acceleration. Again, inference is not illegitimate. Much of science works through inference. But when most of the universe must be explained through unseen components, the confidence of the visible explanation deserves honest examination.

A model can still be mathematically successful and conceptually incomplete.

This is the delicate point.

Modern cosmology has achieved extraordinary things. It has organized enormous data. It has predicted patterns. It has created a working language for cosmic observation. But success in organizing data does not guarantee that the physical interpretation is final. A model may calculate well within its assumptions while still misunderstanding what those assumptions represent.

The discomfort begins when every new sharp observation demands another adjustment.

Too bright? Improve star formation efficiency.

Too massive? Add unusual formation conditions.

Too early? Introduce special seeds.

Expansion mismatch? Modify hidden parameters.

Unexpected structure? Strengthen the role of unseen scaffolding.

This is where the reader must pause.

The universe is not asking for permission to exist. It is not trying to fit itself into our cosmic schedule. If structures appear where we did not expect them, perhaps the first question should not be, "How can we force them into the timeline?" The first question should be, "What assumption made them impossible in the first place?"

That question changes everything.

Because many cosmic tensions arise not merely from data, but from the interpretation imposed upon data. We observe light, redshift, luminosity, motion, distribution, and background signals. Then we place these observations inside a model of expansion, chronology, particle aggregation, spacetime geometry, and hidden components. The anomaly is born at the point where observation refuses that placement.

This means the anomaly is not always in the sky.

Sometimes the anomaly is in the reading.

And this is why modern cosmology has entered an uncomfortable stage. Its instruments are becoming more powerful than its inherited interpretive habits. The sky is no longer vague enough to protect the old comfort. The universe is being seen with sharper eyes, and sharper eyes are less forgiving.

The strange thing is not that the cosmos is producing exceptions.

The strange thing is that we keep calling them exceptions even when they begin forming a pattern.

1.3 — The Patchwork Habit of Modern Cosmology

When a model begins to face pressure, the first response is usually not rejection.

It should not be.

No serious intellectual system should be abandoned at the first sign of difficulty. Science grows by correction, refinement, adjustment, and deeper testing. A theory that has explained many things deserves patience. A model that has organized decades of observations deserves respect. It would be childish to throw away a framework merely because one new observation appears uncomfortable.

But patience is not the same as blind loyalty.

There is a difference between refining a model and rescuing it.

Modern cosmology has slowly entered this delicate zone. Whenever observation refuses the expected script, a new supporting explanation is introduced. Sometimes this may be valid. Sometimes it may be necessary. But when the number of supporting explanations keeps increasing, one must ask whether the original structure is still standing by its own strength, or whether it is being held upright by scaffolding.

The problem is not with one patch.

The problem is the habit of patching.

When galaxies appear too mature at high redshift, the response is to increase the efficiency of early star formation. When they appear too bright, the response is to imagine unusual stellar populations. When supermassive black holes appear too early, the response is to introduce heavy seeds, direct-collapse scenarios, or unusually rapid accretion. When galaxies rotate in ways not explained by visible matter, dark matter becomes necessary. When cosmic expansion appears to accelerate, dark energy enters the stage. When the Hubble tension refuses to disappear, new physics is quietly invited into the waiting room.

Again, none of these ideas should be dismissed casually.

They are not foolish ideas. They are serious attempts by intelligent people to preserve explanatory continuity. Scientists are not sitting in some dark room inventing fantasy. They are responding to real data using the tools and language available within their framework.

But the deeper issue remains.

How many invisible supports can a model require before we admit that the visible structure may be incomplete?

This question is not an insult to science. It is the beginning of honest science.

A patch becomes useful when it improves clarity. It becomes dangerous when it protects an assumption from being questioned. If a new idea helps us see reality more accurately, it is a refinement. If it merely helps the old model survive another contradiction, it becomes intellectual maintenance.

There is a difference.

In ordinary life, we understand this easily. If a house develops one crack, we repair it. If another crack appears, we inspect the wall. If cracks appear across many rooms, near the foundation, across the ceiling, around the pillars, and through the floor, we do not keep applying decorative paint. We call an engineer. We ask whether the structure itself is under stress.

Modern cosmology has reached a similar moment.

The cracks are not in one corner. They appear in cosmic expansion, galaxy formation, black hole growth, matter distribution, gravitational behavior, and the interpretation of cosmic time. The cracks may not mean the entire building must be demolished. But they certainly mean the foundation deserves inspection.

The difficulty is that scientific communities, like all human systems, develop attachment to their working architecture. Once a model becomes dominant, it does not remain only a model. It becomes a language, a career path, a teaching structure, a funding ecosystem, a textbook tradition, and a psychological comfort zone. Thousands of papers are written inside it. Generations of students are trained through it. Instruments are designed according to its expectations. Data is interpreted through its vocabulary.

At that stage, questioning the model feels like questioning the entire intellectual city built around it.

So the system naturally prefers patches.

Patches are safer. They do not disturb the architecture too much. They allow continuity. They let researchers say, "The model still works, but we need this additional component." This is understandable. No civilization of knowledge likes sudden demolition. But reality has no emotional obligation to protect our professional arrangements.

The universe does not care whether our theories are convenient.

It only behaves.

And when its behavior repeatedly requires exceptions, special conditions, hidden components, and mathematical adjustments, the honest mind must become alert.

The most dangerous patch is not the one openly declared.

The most dangerous patch is the hidden assumption beneath all patches: that the core interpretation must be right, and therefore the observation must be made compatible with it.

This is where scientific humility becomes essential.

A model is not reality. It is a disciplined attempt to read reality. Even the most successful model is still a map. And when the map repeatedly demands that the landscape be adjusted in imagination, the problem may be with the map.

Universal Energy Dynamics enters this discussion precisely here.

It does not enter as another patch. It does not say, "Keep the old interpretation and add one more invisible parameter." It asks whether the basic reading itself needs to change. Instead of beginning with particles moving through expanding spacetime, UED begins with structured energy, field tension, ripple behavior, equilibrium-seeking dynamics, and consequences. Instead of forcing observations into a chronological movie reel, it asks whether cosmic depth may be field depth, whether redshift may carry structural strain, and whether so-called anomalies are signals of misread field behavior.

This is a different move.

It does not repair the old wall.

It examines the foundation.

And that is why the patchwork habit matters. It shows us where the old model is struggling most. Every patch is a clue. Every exception is a signal. Every tension is a place where reality is pressing against interpretation.

The purpose is not to ridicule the patch.

The purpose is to read what the need for the patch is silently telling us.

Because at some point, if a model survives only by continuously adding exceptions, the anomaly is no longer outside the model.

The anomaly is the model's dependence on exceptions.

1.4 — The Hidden Assumption of Cosmic Chronology

The deepest assumptions are often the least visible.

They do not stand in front of us like declared theories. They sit quietly behind the way we think. They guide our questions before we even realize we are asking them. They decide what will appear normal and what will appear impossible. They decide which observations will be welcomed as confirmation and which will be pushed into the category of anomaly.

In modern cosmology, one such hidden assumption is cosmic chronology.

We have been trained to believe that looking farther into space means looking backward into time in a simple, direct, and absolute way. This idea has become so familiar that most people no longer feel the need to question it. The statement sounds almost natural: because light takes time to travel, distant light must show us the past. Therefore, the deeper we look, the earlier the universe must be.

At one level, this is reasonable.

Light does take time to reach us. Observation is not instantaneous. A star seen from far away is not seen as it is at this exact local moment. There is delay. There is transmission. There is measurement across distance. No serious reinterpretation needs to deny this basic observational fact.

But the problem begins when this observational delay is converted into a complete cosmic story.

From "light takes time to arrive," we jump to "we are seeing the universe exactly as it was in its historical childhood." From "high redshift indicates a distant observational condition," we jump to "this object belongs to an early immature stage of universal evolution." From "the signal has travelled across a large cosmic interval," we jump to "the structure must be young, primitive, and still under development."

This jump is not purely observational.

It is interpretive.

And once this interpretation becomes dominant, every distant object is forced into a historical role. A high-redshift galaxy is not merely observed; it is assigned an age-based identity. It becomes an "early galaxy." A bright compact structure becomes a problem because brightness and compactness are not expected in that chapter of the cosmic timeline. A massive black hole becomes impossible because there seems to be insufficient time for it to grow inside the assigned historical window.

The anomaly is born from the assignment.

The galaxy itself does not say, "I am too early."

The black hole candidate does not say, "I grew too fast."

The redshift does not say, "Interpret me only as chronological distance."

These meanings are added by the model.

This is where Universal Energy Dynamics begins to ask a different question. What if cosmic depth is not only cosmic past? What if deep observation is also field penetration? What if redshift carries not only the story of distance and expansion, but also the signature of field strain, resistance, density of interaction, and state-dependent energy behavior? What if what we call the early universe is not simply an ancient photograph, but a deeper layer of the present universe being resolved through advanced observation?

This does not mean every existing measurement is useless.

It means the meaning of measurement may need refinement.

When JWST observes high-redshift regions, the mainstream reading immediately places them inside the early-universe timeline. UED asks whether those regions may instead represent high-tension layers of the Structured Vacuum Energy continuum. Their luminosity, compactness, and apparent maturity may not be signs of impossible speed in the past. They may be signs of intense field realignment in the present structure of the cosmos.

This shift is subtle but powerful.

In the old reading, the universe is a movie reel. We sit at one point and look backward into earlier frames. The farther we look, the earlier the frame. The earlier the frame, the more immature the structure should be.

In the UED reading, the universe is not primarily a movie reel. It is a structured field-depth. Observation does not merely retrieve old frames. It interacts with layers of energy configuration, resistance, ripple condition, and alignment delay. What appears distant may also be structurally different. What appears early may also be field-dense. What appears ancient may be a present-layer phenomenon filtered through observational delay and interpreted through chronological habit.

This matters because the movie-reel model creates expectations that may not belong to reality.

It expects infancy at depth.

It expects simplicity before complexity.

It expects slow mechanical assembly.

It expects history to behave like a human biography.

But the universe is not a human biography.

It may not require the kind of childhood we imagine. Structure may not be produced only by slow accumulation. Organization may not be merely the reward of duration. A field-based universe may generate structure wherever conditions reach the required threshold of alignment. In such a universe, maturity is not always a function of age. It may be a function of field condition.

This is why cosmic chronology must be examined carefully.

The issue is not whether observational delay exists. It does. The issue is whether delay alone justifies the complete historical interpretation attached to distant observations. The issue is whether we have mistaken signal travel time for universal autobiography. The issue is whether the universe is being read as a timeline because the human mind loves timelines.

Human consciousness arranges experience through before and after. We remember, compare, predict, and narrate. This is useful for life, but dangerous when projected onto the whole cosmos without sufficient caution. The mind naturally converts depth into history, sequence into meaning, and delay into story.

But reality may be deeper than narration.

The cosmos may not be preserving old scenes for us like a museum of dead light. It may be presenting layered structural conditions that our instruments receive and our models translate into history. The translation may be useful in some cases, but incomplete in others. And when the translation fails, we call the result an anomaly.

This is the hidden assumption we must bring into daylight.

Perhaps the universe is not strange.

Perhaps the timeline is too strict.

Perhaps JWST is not humiliating the cosmos but humiliating our habit of turning every distant signal into a chapter of a pre-written story.

The question, then, is not only how far we are looking.

The question is: what kind of reality are we assuming when we interpret what we see?

1.5 — Anomaly as a Message from Reality

An anomaly is not the enemy of science.

In fact, anomaly is one of the most honest friends science can have.

A comfortable observation confirms what we already believe. It gives confidence, but it does not always deepen intelligence. A difficult observation disturbs us. It forces us to look again. It exposes the weak joints of our explanation. It asks whether the model is reading reality or merely defending itself.

That is why anomaly must not be treated as embarrassment.

It must be treated as communication.

When reality does not fit our expectation, reality is not misbehaving. It is speaking in a language our current framework may not yet understand. The problem is not that the universe has violated the rule. The problem may be that the rule was never wide enough to include the real behavior of the universe.

This is a difficult lesson because human intelligence becomes attached to its own maps.

Once a model becomes successful, we begin to trust it deeply. We stop seeing it as a temporary reading and begin to treat it as reality itself. Then, when observation disagrees with the model, we feel as if reality has become strange. But reality has not changed its nature. Only our expectation has been disturbed.

A map is useful.

But a map is not the territory.

If the map shows a straight road and the traveler finds a mountain, the mountain is not an anomaly. The map is incomplete. If a child draws the sky as a blue flat sheet and later sees clouds, stars, storms, and lightning, the sky has not betrayed the child. The child's drawing was too simple. The same thing happens in science, but with more equations, more confidence, and more institutional weight.

When JWST observes galaxies that appear too mature for the expected cosmic timeline, the galaxies are not guilty. They are not violating an oath given to cosmologists. They are simply existing according to their own structural reality. The discomfort arises because our interpretive framework has assigned them a role they refuse to play.

We call them "too early."

But too early according to what?

According to the timeline.

We call them "too massive."

But too massive according to what?

According to the assumed growth pathway.

We call black holes "too developed."

But too developed according to what?

According to the idea that such objects must grow slowly through chronological accumulation.

The anomaly does not begin in the object.

It begins in the comparison between observation and assumption.

This does not mean every anomaly immediately proves a new theory. That would be irresponsible. Many anomalies are resolved through better data, better instruments, or better understanding of local conditions. Science must remain disciplined. But when anomalies become patterned, repeated, and structurally connected, they deserve more than technical patchwork. They deserve philosophical honesty.

The question must shift.

Instead of asking only, "How do we explain this within the existing model?" we must also ask, "What assumption made this observation appear impossible?"

That second question is more dangerous.

It is also more powerful.

Because it does not simply protect the old framework. It examines the frame itself. It asks whether our categories are too narrow. It asks whether our language has become outdated. It asks whether our imagination has become trapped inside the success of earlier models.

This is where Universal Energy Dynamics finds its opening.

UED treats anomalies not as random defects, but as stress signals between reality and interpretation. A cosmic anomaly is not necessarily a broken piece of the universe. It may be a point where the field is expressing behavior that particle-based, spacetime-based, chronology-heavy interpretation cannot fully absorb.

In that sense, anomaly becomes evidence of interpretive pressure.

A bright high-redshift galaxy may not be a structure that grew too fast in the past. It may be a high-density field consequence being misread through chronological expectation. A supermassive black hole candidate may not be an object that mysteriously skipped its childhood. It may be a central regulatory saturation zone established with the galactic structure itself. Hubble tension may not be a crisis in expansion rate alone. It may be a sign that different field states produce different alignment delays.

The anomaly is the doorway.

But only if we do not immediately close it with familiar language.

Reality often teaches through discomfort. In personal life, a repeated problem is rarely just a problem; it is a signal. If the same conflict returns again and again, something deeper is misaligned. If the same stress appears in different situations, the issue is not only the situation; it is the pattern beneath. If the same crack appears after every repair, the wall is not the final issue; the foundation must be examined.

Cosmology is not separate from this logic.

When galaxies, black holes, redshift, expansion, dark matter, dark energy, and cosmic maturity repeatedly demand special explanations, the honest mind must pause. Perhaps these are not scattered puzzles. Perhaps they are messages from a deeper structure of reality.

The universe may be saying: do not only adjust the numbers; reconsider the lens.

This is why anomaly should be respected.

An anomaly is not the death of knowledge. It is the beginning of a more mature form of knowledge. It tells us that reality is larger than our present explanation. It reminds us that intelligence must remain alive, not frozen in admiration of its own previous success.

The tragedy begins when science becomes afraid of its own anomalies.

Because anomaly is where discovery breathes.

If every uncomfortable observation is immediately forced into old language, then science becomes a management system, not a search for truth. It begins maintaining order instead of expanding understanding.

It becomes more concerned with protecting continuity than confronting reality.

But reality does not need our continuity.

Reality only needs to be seen as clearly as possible.

So when this book speaks of the end of cosmic anomalies, it does not mean the end of unknowns. It does not mean the universe will become simple, dull, or fully explained. It means that many so-called anomalies may stop looking anomalous once we stop forcing them into the wrong interpretive frame.

The cosmos may still be mysterious.

But it may not be confused.

The confusion may belong to our reading.

An anomaly is reality asking intelligence to become more honest. It is the moment when the universe refuses flattery and demands maturity. It is not a cosmic mistake. It is a signal at the boundary of an old understanding.

And if we are humble enough, that boundary can become the birthplace of a new science.

1.6 — Why UED Enters the Discussion

Universal Energy Dynamics does not enter this discussion as a decorative alternative.

It does not arrive merely to rename old problems in new language. It does not say that mainstream cosmology has observed nothing valuable. It does not dismiss decades of scientific effort as useless. That would be careless, and carelessness is not science.

UED enters because the pressure points in modern cosmology are no longer small. They are conceptual. They are not limited to one strange galaxy, one strange black hole candidate, or one unresolved measurement. They touch the way we understand matter, gravity, time, distance, observation, structure, and cosmic evolution itself.

When the discomfort is this deep, another parameter is not enough.

Another patch is not enough.

Another invisible helper may not be enough.

At some stage, the lens itself must be examined.

This is where UED begins.

The standard reading of the universe often begins with separate things: particles, objects, forces, spacetime, expansion, and chronological development. It imagines reality as something built from parts over time. The universe begins, expands, cools, forms particles, gathers matter, builds stars, builds galaxies, and then gradually becomes complex.

This is a construction-based imagination.

UED begins differently.

It begins with the idea that existence is not a pile of independent things waiting to be assembled. Existence is a continuous field consequence. What we call matter, motion, gravity, time, light, and structure are not isolated fundamentals. They are expressions of deeper energy behavior. They arise from structured energy, field interaction, ripple response, alignment, resistance, and the universal tendency toward equilibrium.

In simple words, UED does not ask first, "What objects are moving?"

It asks, "What field condition is producing this observable consequence?"

This shift changes everything.

A galaxy is no longer merely a collection of stars slowly assembled through mechanical accumulation. It becomes a large-scale field consequence, a structured expression of energy alignment within the deeper vacuum medium. A black hole is no longer treated as an impossible object of infinite density. It becomes a saturation region, a regulatory threshold where ordinary ripple coherence collapses from the standpoint of external observation. Gravity is no longer a mysterious pull or only the curvature of a mathematical fabric. It becomes an emergent consequence of energy gradients seeking balance.

Even time changes.

In the usual imagination, time is the invisible road on which the universe travels. Events happen "in" time. Objects evolve "through" time. Distant galaxies belong to earlier "chapters" of the cosmic story.

But UED treats time more carefully. Time is not taken as an independent ruler standing outside reality. It is understood as Alignment Threshold Time: the delay required for energy states to rearrange, interact, and become measurable as consequence. Time is not denied, but its status changes. It becomes a function of causal resolution, not a cosmic road.

This matters greatly when we interpret deep-space observations.

If time is treated as an absolute cosmic highway, then distant observations are immediately placed into a historical sequence. But if time is connected to alignment delay and field-state interaction, then high-redshift observation may not be merely a window into primitive history. It may also be a reading of high-resistance field layers, deeper structural conditions, and intense ripple tension within the present universe.

This is why UED does not see JWST's observations as shocking in the same way.

Where mainstream cosmology sees mature galaxies too early, UED asks whether structure may emerge wherever field conditions reach the necessary alignment threshold. Where mainstream cosmology sees supermassive black holes that grew impossibly fast, UED asks whether central regulators form together with galactic structures as part of the same field-realignment event. Where mainstream cosmology sees Hubble tension as a crisis in expansion measurement, UED asks whether different field states naturally create different alignment delays.

The language changes because the lens changes.

This is important.

New words are not enough. Any theory can invent terms. The value of a framework depends on whether its terms reveal a clearer relationship between observation and mechanism. In UED, terms such as Structured Vacuum Energy, ripple alignment, field tension, Ripple Saturation Zone, and Alignment Threshold Time are not ornamental. They are attempts to describe a universe that behaves less like a machine assembled from dead parts and more like a structured field continuously resolving its own internal disturbances.

This does not make the universe magical.

It makes it more coherent.

The word "energy" is often used casually, even carelessly. But in UED, energy is not a vague spiritual decoration. It is the operational basis of manifestation. It is the actionable expression through which reality becomes structured, measurable, and consequential. What appears as matter is energy held in structure. What appears as force is energy imbalance seeking resolution. What appears as gravity is gradient behavior. What appears as time is the delay of alignment. What appears as anomaly is the friction between reality and incomplete interpretation.

This is why UED must enter the discussion now.

Because modern cosmology is not only facing new data. It is facing the consequences of its own starting points. If one begins with particles, spacetime, expansion, and chronology, then every observation must be squeezed into that grammar. But if the universe is more fundamentally a structured energy field, then the same observations may carry different meaning.

The mature galaxy is no longer an embarrassment.

The early black hole is no longer an impossibility.

The Hubble tension is no longer only a crisis.

The cosmic timeline is no longer the unquestioned judge.

They become signals of field behavior.

This does not mean UED has answered everything. No honest framework should claim final completion before sufficient testing, refinement, and criticism. But UED offers something important: it changes the question. It allows us to stop asking only how to protect the old story and begin asking how reality may actually be operating beneath the story.

That is the real reason UED enters this book.

Not to decorate cosmology.

Not to rebel for drama.

Not to win an argument by loud rejection.

UED enters because the anomalies are asking for a deeper grammar.

They are asking for a framework in which structure does not depend only on chronological age, gravity does not depend only on curved spacetime, black holes do not require infinite absurdities, and time does not rule the universe like an invisible god.

The old language calls these observations strange.

UED asks whether they are strange only because the language is old.

And once that question is allowed, the entire cosmic conversation changes.

Closing Reflection — The Cosmos Was Not Waiting for Our Permission

The universe was not waiting for our permission to behave as it does.

It did not design galaxies according to our timelines. It did not form structures according to our academic comfort. It did not create black holes, redshift, expansion, or cosmic depth so that human theories could remain undisturbed. The cosmos has always been what it is. Only our ability to observe it has changed.

And when observation becomes sharper, illusion becomes more difficult to maintain.

This is the quiet lesson of modern cosmology.

For many years, the standard cosmic story gave us confidence. It arranged the universe into a timeline. It gave every structure a place, every event a sequence, every distance a historical meaning. It made the universe intellectually manageable. It allowed human intelligence to feel that the grandest reality could be narrated like a disciplined biography.

But the universe is not a biography.

It is not obligated to have childhood, youth, maturity, and old age in the way the human mind expects. It is not required to build slowly only because our models need duration. It is not forced to place simplicity before complexity merely because our thinking prefers linear development.

This is where the discomfort begins.

When a high-redshift galaxy appears too mature, we call it strange. When a supermassive black hole candidate appears too early, we call it a puzzle. When the Hubble tension persists, we call it a crisis. When invisible components are required to preserve visible behavior, we call them necessary.

But perhaps all these labels reveal more about us than about the universe.

Perhaps "too early" only means too early for our timeline.

Perhaps "too massive" only means too massive for our growth story.

Perhaps "too strange" only means too strange for our inherited language.

The cosmos itself does not carry these embarrassments. It does not apologize for its structures. It does not explain itself in the vocabulary of our textbooks. It simply exists, moves, balances, condenses, saturates, redistributes, and resolves according to deeper principles that may not yet be fully captured by mainstream interpretation.

That is why this chapter began with a simple statement: the universe did not become strange overnight.

It only became harder to misread.

JWST did not create cosmic anomalies. It made our old assumptions visible. It showed that the gap between observation and interpretation may be wider than we wanted to admit. It revealed that a model can remain useful and still be incomplete. It reminded us that successful calculation is not always final understanding.

This does not mean we must throw away all existing science.

That would be reckless.

Every model carries the effort of human intelligence. Every theory is a milestone. Every equation is a disciplined attempt to touch reality. The standard model of cosmology has helped organize enormous observation and has given humanity a powerful working structure.

But no model deserves permanent immunity.

A theory must serve reality. Reality must not be forced to serve theory.

When repeated anomalies appear, mature intelligence does not panic, and it does not blindly defend. It pauses. It asks what the anomalies are trying to reveal. It investigates whether the fault lies in the data, the instrument, the parameter, or the interpretation itself.

This pause is where a new science begins.

Universal Energy Dynamics enters through this pause. It does not enter as another patch placed upon the old wall. It enters as a different way of looking at the wall, the crack, the foundation, and the ground beneath. It asks whether matter is better understood as field consequence. It asks whether gravity is better understood as energy-gradient behavior. It asks whether black holes are better understood as ripple saturation zones. It asks whether time is not a cosmic road but an alignment delay. It asks whether redshift may carry signatures of field strain, not merely chronological distance.

These questions may disturb established comfort.

Good.

A question that does not disturb anything rarely discovers anything.

The purpose of this book is not to make the universe smaller, simpler, or conveniently explained. The purpose is to make our interpretation more honest. Mystery will remain. Unknowns will remain. But confusion created by wrong framing can be reduced.

The end of cosmic anomalies does not mean the end of cosmic wonder.

It means the end of blaming the cosmos for the limitations of our reading.

The universe was never broken. It was never late. It was never premature. It was never confused. It was never violating the script.

Perhaps there was no script.

Perhaps there was only reality, and human intelligence kept writing stories around it.

Now, sharper observation has begun to challenge those stories. And if we are courageous enough, we may stop asking why the universe is behaving strangely and begin asking why we expected it to behave otherwise.

That is the real beginning of this book.

Not the end of mystery.

The end of unnecessary anomaly.

Chapter 2 — JWST and the Collapse of the Cosmic Timeline

Why the James Webb Space Telescope may not be showing us the "early universe" in the way mainstream cosmology assumes.

The Telescope That Disturbed the Story

Every powerful instrument changes the world twice.

First, it changes what we are able to see.

Then, more dangerously, it changes what we are no longer able to believe comfortably.

The James Webb Space Telescope belongs to this second category. It is not only a telescope. It is a disturbance. It is a silent question placed in front of modern cosmology. It did not shout. It did not argue. It simply looked deeper, clearer, and with greater sensitivity than earlier instruments. And by doing that, it quietly disturbed one of the most trusted stories of modern science.

We expected confirmation.

We received discomfort.

That is the interesting part.

Humanity did not build JWST merely as a beautiful machine floating in space. It was built with expectation. It was built with confidence. It was built as an advanced eye that would look toward what mainstream cosmology calls the cosmic dawn. It was supposed to take us closer to the earliest chapters of the universe. It was supposed to show us primitive galaxies, early stars, incomplete structures, and the slow beginning of cosmic architecture.

In simple words, JWST was expected to sharpen the old timeline.

But sharper vision does not always obey expectation.

When the telescope began revealing deep-field structures, the picture did not remain as obedient as expected. Galaxies appeared brighter. Some appeared more massive. Some seemed too organized. Some compact red objects looked too mature for the age assigned to them. Possible early black hole candidates appeared in places where the standard growth story becomes uncomfortable.

Naturally, the language of surprise began.

Too early.

Too bright.

Too massive.

Too mature.

Too fast.

These phrases sound scientific, but they carry a hidden confession. They do not describe the universe alone. They describe the mismatch between the universe and the story we had already written about it.

A galaxy is not "too early" by itself.

It becomes too early only when a timeline has already decided what early should look like.

A structure is not "too mature" by itself.

It becomes too mature only when our model says maturity must require a certain amount of chronological development.

This is where JWST becomes more than an observational instrument. It becomes a philosophical instrument. It forces us to ask whether we are truly seeing the universe, or whether we are seeing our assumptions become uncomfortable in front of the universe.

The common explanation is that JWST is looking into the past. The deeper it looks, the earlier it sees. The earlier it sees, the younger the universe must be. This idea is repeated so often that it begins to feel unquestionable. It has become part of the popular imagination: telescopes are time machines, deep space is ancient history, and distant light is a cosmic archive.

At one level, this is understandable. Light takes time to travel. Observation involves delay. A signal reaching us from far away does not belong to our immediate local moment. This is not the issue.

The issue begins when signal delay becomes cosmic biography.

From the fact that light travels, we build the idea that we are watching the universe's childhood exactly as it was. From redshift, we build chronological certainty. From distance, we build history. From history, we build expectation. Then, when observation refuses that expectation, we call reality anomalous.

But what if the telescope is not wrong?

What if the data is not confused?

What if the problem lies in the way we convert cosmic depth into cosmic past?

This chapter begins there.

JWST may not be showing us the "early universe" in the simple way mainstream cosmology assumes. It may be showing us deeper structural layers of the present universe. It may be penetrating regions of high field resistance, high ripple tension, and compact energy configuration that older instruments could not resolve. It may be revealing not dead historical frames, but living depth within the continuous Now.

This is not a small shift.

It changes the metaphor.

The universe may not be a movie reel where each distant object represents an earlier frame in a completed film. It may be a structured field-depth where observation unfolds layers according to sensitivity, resistance, alignment delay, and energy condition.

If this is true, then JWST has not collapsed the universe.

It has collapsed the comfort of the cosmic timeline.

And perhaps that is exactly why it matters.

2.1 — What JWST Was Expected to Confirm

The James Webb Space Telescope was not launched into an empty intellectual space.

It carried a machine into space, but it also carried a story.

Behind its mirrors, instruments, sensors, cooling systems, and engineering brilliance, there was an expectation. Humanity had already built a broad cosmic timeline, and JWST was expected to look deeper into that timeline with sharper eyes. It was expected to take the accepted story of the universe and give it better photographs, better details, better evidence, and better confidence.

That expectation was natural.

Every great scientific instrument is built not only to discover the unknown, but also to test what we think we already know. We do not build telescopes as innocent eyes. We build them inside frameworks. We build them with assumptions, models, predictions, and questions already loaded into our imagination. The instrument looks outward, but the mind behind the instrument already carries a map.

JWST was expected to look toward what modern cosmology calls the cosmic dawn.

That phrase itself is powerful. Cosmic dawn suggests the first light after darkness, the earliest formation of stars, the first galaxies assembling from simpler beginnings, the first glow of structure emerging from a young universe. It suggests infancy. It suggests beginning. It suggests a time when the universe was still learning to organize itself.

In the mainstream story, this makes sense.

After the Big Bang, the universe expands and cools. Matter begins to form. Atoms emerge. Gravity slowly gathers matter into clouds. Stars ignite. Galaxies begin to assemble. Structure grows gradually through attraction, collapse, collision, merger, and accumulation. Complexity arrives step by step. The early universe, therefore, should not look fully mature. It should look like a workshop under construction.

JWST was expected to enter that workshop.

It was expected to show the unfinished stage of cosmic architecture. It was expected to reveal faint early galaxies, primitive stellar populations, irregular shapes, small structures, and the slow beginning of large-scale organization. The deeper it looked, the more primitive the structures were expected to become. This was the natural prediction of the timeline.

The logic appeared simple.

If farther means older, and older means earlier, then earlier should mean less developed.

That was the comfort.

And this comfort was not foolish. It had observational and mathematical backing within the accepted framework. Earlier telescopes had already given humanity glimpses of distant galaxies and deep fields. The cosmic microwave background had been interpreted as a remnant of the early universe. Redshift had been used as a key marker of distance and expansion. The standard model had organized these signals into a working cosmic history.

So JWST was not expected to destroy the story.

It was expected to refine it.

It was expected to show the first chapters more clearly. It was expected to help scientists understand how the first galaxies formed, how early stars behaved, how black holes began, and how structure gradually emerged. The mission was celebrated because it promised to take us closer to the beginning—not necessarily to question whether the beginning was being interpreted correctly.

This is the subtle point.

JWST was not merely expected to observe. It was expected to observe within a pre-written cosmic grammar.

That grammar said deep space is early time.

It said early time means primitive structure.

It said maturity requires duration.

It said massive galaxies need long growth.

It said supermassive black holes need a chronological path.

It said the universe moves from simplicity to complexity in a broad evolutionary sequence.

Inside this grammar, JWST had a clear role. It was supposed to be the best reader of the earliest pages.

But what happens when the reader opens the early pages and finds later chapters already written there?

That is where discomfort begins.

Because if JWST had found only what the model expected, the telescope would have been celebrated as a magnificent confirmation. It would have strengthened the timeline. It would have allowed modern cosmology to say, "Look, the early universe behaves as predicted. We are seeing the childhood of cosmic structure."

But the universe did not remain that obedient.

The expectation of faintness, immaturity, and slow assembly was disturbed by observations that appeared too bright, too compact, too massive, or too organized. The telescope began showing not only beginnings, but structure. Not only infancy, but unexpected maturity. Not only faint seeds, but surprising density of cosmic formation.

This does not automatically mean every mainstream interpretation is wrong.

But it does mean the expectation must be examined.

The most important thing about JWST is not only what it saw. It is what we expected it to see before it saw anything.

Expectation decides surprise.

If the mind expects a child and sees an adult, it calls the adult impossible. But perhaps the problem is not the adult. Perhaps the problem is that the mind entered the room with the wrong assumption.

This is exactly the difficulty with the cosmic timeline.

JWST was expected to confirm the idea that deeper observation means earlier cosmic childhood. But its observations forced a harder question: what if our idea of childhood belongs more to human narration than to the universe itself?

The telescope did not fail.

The data did not betray science.

The discomfort emerged because the universe appeared less willing to behave like a slow mechanical construction project than the timeline had suggested.

And that is why JWST matters so deeply for this book.

It did not merely look farther.

It tested the foundation of "farther."

2.2 — The Shock of Mature Structures at Cosmic Depth

The real shock of JWST was not that it saw more.

The shock was that it saw more than the timeline was emotionally prepared to accept.

A telescope with greater power is expected to reveal hidden details. That much was obvious. Better resolution should uncover fainter galaxies, deeper signals, more distant structures, and previously unseen cosmic formations. No one expected JWST to see nothing unusual. Science always expects surprise at the edge of observation.

But this surprise was different.

It did not merely add new objects to the existing story. It disturbed the story's sequence.

When JWST began resolving extremely distant regions, scientists expected to find the early universe behaving like an early universe. That means young structures, faint beginnings, rough formations, incomplete organization, and galaxies still struggling to assemble themselves. In the standard imagination, deep cosmic observation should reveal the childhood of structure.

Instead, in many cases, the picture looked uncomfortably mature.

Some galaxies appeared brighter than expected. Some appeared more compact. Some appeared more massive. Some seemed to have developed too much structure within the limited time available under the accepted cosmic timeline. Some compact red objects, popularly discussed as "Little Red Dots," seemed to carry an intensity and density that did not sit easily inside the usual slow-growth expectation.

This created a language of discomfort.

Too early.

Too bright.

Too massive.

Too mature.

Too well formed.

These phrases are important because they do not belong purely to observation. They belong to comparison. The object is observed, but the judgment arises only when the object is compared with the expected model.

A galaxy does not announce, "I am too early."

It simply appears.

A structure does not say, "I am too massive for my age."

It simply presents its light, redshift, brightness, shape, and energetic signature.

The phrase "too early" is born inside the model, not inside the galaxy.

This is the key point.

If a model says that a certain kind of structure should require a long duration to form, then any such structure appearing at great cosmic depth becomes a problem. But the problem is not the structure alone. The problem is the assumption that maturity must always require the duration imagined by the model.

In ordinary life, we understand this easily. If we believe a child of ten cannot play complex music, and then we see a ten-year-old perform beautifully, we may say, "This is impossible." But the performance is not impossible. Our assumption was incomplete. Perhaps the child had unusual training. Perhaps talent, environment, and intensity changed the expected development. Perhaps our timeline of ability was too rigid.

In cosmology, the same logic becomes much larger.

If the universe is imagined as a slow mechanical construction project, then maturity at depth becomes shocking. If galaxies must assemble gradually through collisions, accretion, star formation, dark matter scaffolding, and long cosmic evolution, then mature-looking structures at high redshift appear to break the expected sequence.

But what if the sequence itself is incomplete?

What if structure does not always require the kind of chronological childhood we have assigned to it?

What if the field condition matters more than the age assigned by the timeline?

Under Universal Energy Dynamics, this discomfort is not surprising. A galaxy is not merely a pile of matter slowly gathered through mechanical events. It is a field consequence. It is a structured expression of energy alignment, ripple condensation, and equilibrium-seeking behavior within the deeper medium of Structured Vacuum Energy.

In this reading, cosmic maturity is not only a function of chronological age. It is also a function of field intensity, energy density, ripple saturation, local equilibrium conditions, and structural alignment. Wherever the field reaches the necessary threshold, structure may appear far more rapidly than a particle-aggregation timeline would comfortably allow.

This does not mean the observations are false.

It means their meaning may be different.

The mainstream interpretation says: "These galaxies are too mature for the early universe."

UED asks: "Why are we assuming that these regions are early in the same simple chronological sense?"

The mainstream interpretation says: "These structures formed too quickly."

UED asks: "What if they are not slow-built objects but field-aligned consequences?"

The mainstream interpretation says: "We must explain how such mass accumulated so fast."

UED asks: "What if mass itself is not an independent substance accumulated from outside, but a structured energy consequence stabilized within the field?"

This is why JWST creates such a powerful turning point. It does not merely give us a few technical puzzles. It forces us to examine the grammar through which we convert light into history, structure into age, and depth into infancy.

The shock is not just observational.

It is interpretive.

If JWST had shown only weak, faint, primitive structures, the standard timeline would have become stronger. The telescope would have been celebrated as the great confirmer of cosmic childhood. But when it showed signs of unexpected maturity, the timeline itself had to defend its assumptions.

Naturally, the existing framework began searching for solutions.

Perhaps stars formed more efficiently.

Perhaps early galaxies converted gas into stars faster than expected.

Perhaps unusual stellar populations made them brighter.

Perhaps black hole activity contributed to the luminosity.

Perhaps some mass estimates require revision.

Perhaps dust, chemistry, or observational bias affects interpretation.

All of these possibilities deserve investigation. Science must test them carefully. But even while such explanations are explored, the deeper question remains untouched unless we are courageous enough to ask it.

Why did we expect immaturity there in the first place?

This question is not a technical detail. It is the doorway to a new interpretation.

Because the word "mature" itself belongs to a timeline. It assumes growth, duration, stages, infancy, development, and completion. But if the universe is better understood as a structured field resolving conditions rather than a historical machine assembling parts, then maturity may not mean what we think it means.

A field can organize when the condition is right.

A ripple can stabilize when equilibrium demands it.

A structure can emerge when tension reaches the point of alignment.

A region can become luminous not because it has lived long, but because it is energetically intense.

This is the UED shift.

JWST's mature structures at cosmic depth may not be impossible old objects appearing in a young universe. They may be high-tension field consequences appearing through a deeper observational layer of the present universe. Their brightness may not be merely a sign of ancient star populations. Their compactness may not be merely a sign of accelerated mechanical growth. Their apparent maturity may be a sign of strong field alignment.

Once this possibility is allowed, the shock begins to soften.

Not because the universe becomes less mysterious, but because the anomaly becomes less artificial.

The cosmos may not be violating the timeline.

The timeline may be violating the cosmos by forcing it into a human sequence.

This is why the mature structures seen by JWST matter so much. They do not only challenge a few numbers. They challenge the habit of reading the universe as if it must grow like a plant, age like a body, and develop like a civilization.

The universe may operate differently.

It may not require our familiar idea of youth before structure.

It may not require our emotional comfort of gradual maturity.

It may not be bound to the slow classroom logic we impose upon it.

And that is the real shock.

JWST did not merely find mature galaxies.

It found the weakness of our expectation.

2.3 — The Problem with Calling Deep Space the Past

One of the most repeated statements in popular cosmology is that telescopes are time machines.

It is a powerful sentence.

It is also emotionally satisfying.

When we hear that a telescope is looking billions of years into the past, the mind immediately becomes excited. It feels as if science has found a way to break the ordinary limits of human experience. We cannot travel back in time, but perhaps we can look back in time. We cannot return to the beginning, but perhaps light has preserved the beginning for us. We cannot witness the childhood of the universe directly, but perhaps deep space is a cosmic archive waiting to be opened.

This idea has poetic force.

It is not entirely baseless either.

Light does take time to travel. A signal coming from a distant object reaches us after a delay. When we observe the Sun, we do not see it in the exact instant of its present condition; we see it after light has taken time to reach Earth. When we observe distant stars and galaxies, that delay becomes enormous. In this basic sense, observation does involve the past.

But here we must slow down.

A delay in signal arrival is one thing.

A complete historical interpretation of the universe is another.

This difference is extremely important.

The fact that light takes time to travel does not automatically mean that every deep observation must be treated as a simple photograph from a neat cosmic childhood. The delay tells us something about transmission. It does not, by itself, guarantee that our full story about cosmic development, maturity, distance, expansion, and age is complete.

Yet modern cosmology often moves too quickly from signal delay to historical certainty.

A distant galaxy is observed.

Its redshift is measured.

Its light is interpreted through the expansion model.

Its place is assigned inside the cosmic timeline.

Then its structure is judged according to what that timeline permits.

If the galaxy appears small and faint, it comfortably fits the expected early universe. If it appears bright, dense, organized, or massive, discomfort begins. The object is then called surprising, puzzling, unusually mature, or even impossible for its time.

But again, we must ask: whose time?

The galaxy's time?

Or the timeline assigned by our interpretation?

This is the hidden problem with calling deep space the past. It sounds like a direct statement of observation, but it often carries a heavy package of assumptions. It assumes that the universe is best understood as a sequence of historical frames. It assumes that distance can be converted into age with confidence. It assumes that age determines maturity in a predictable way. It assumes that deeper visibility must reveal earlier developmental stages.

The phrase "looking into the past" becomes more than an observational description.

It becomes a worldview.

And once it becomes a worldview, it quietly controls interpretation.

Universal Energy Dynamics asks us to reopen this issue.

UED does not need to deny light-travel delay. Denying delay would be unnecessary and unhelpful. The question is not whether signals require time to reach an observer. The question is whether our meaning of that delay has become too narrow.

What if deep observation is not only historical retrieval?

What if it is also field penetration?

What if JWST is not merely collecting dead light from dead stages of the universe, but resolving deeper structural layers of the present cosmic field?

This may sound strange only because the old metaphor is so deeply installed in the mind. We have become used to imagining the universe as a film reel. In that film reel, every distant object is an older frame. The telescope becomes a projector running backward. The deeper we look, the closer we come to the first scene.

But perhaps the universe is not a film reel.

Perhaps it is more like an ocean.

When we look deeper into the ocean, we are not looking into the ocean's past. We are looking into deeper layers of the same ocean. Those layers may have different pressure, temperature, density, motion, and life forms. A shallow observer may wrongly imagine that depth means history, but depth is not automatically past. Depth can mean condition.

This analogy is not perfect, but it is useful.

JWST may be doing something similar at the cosmic level. It may be looking into deeper field conditions, not merely earlier historical stages. It may be resolving high-resistance layers of the Structured Vacuum Energy continuum that older instruments could not access. What appears as distant and ancient may also be dense, strained, compact, and energetically intense within the present structure of the cosmos.

This changes the interpretation of surprise.

If deep space is only the past, then mature structures at high redshift create a chronological crisis. They appear too developed for their assigned age. The model must then explain how they formed so quickly.

But if deep observation also means field-depth observation, then the question changes. We no longer ask only, "How did this structure grow so fast long ago?" We also ask, "What present field condition is being read as ancient structure?"

This is not a small shift.

It moves cosmology from chronological storytelling toward structural interpretation.

The mainstream view often treats deep light as a messenger from an ancient moment. UED treats the signal as a consequence of field interaction, alignment delay, resistance, and observational resolution. The signal does not merely carry history. It carries condition. It carries the behavior of energy across the medium. It carries the imprint of the field through which it is observed and from which it emerges.

This is why calling deep space "the past" may be too simplistic.

It is not completely false.

It is incomplete.

And incomplete ideas become dangerous when they are treated as final.

A child looking at an old photograph of a person may say, "This is the person." But the photograph is only one captured condition. It is not the total existence of that person. Similarly, light arriving after a delay may reveal an observational condition, but it may not fully justify all the historical meaning imposed upon it.

The universe may not be preserving itself as a museum of old scenes.

It may not be storing ancient frames for our telescopes to recover.

It may not be saying, "Here is my childhood, please arrange it according to your timeline."

That is our narration.

The cosmos may be offering signals from structural depth, and our mind may be translating those signals into history because the mind is addicted to sequence.

This addiction is understandable. Human life depends on past, present, and future. Memory builds identity. Prediction helps survival. We naturally arrange experience into before and after. So when we look at the cosmos, we carry the same habit. We transform delay into history, history into age, age into expectation, and expectation into judgment.

But the universe may not think like the human mind.

It may not arrange itself in the same way we arrange personal experience.

From the UED perspective, the universe is a continuous field of structured energy seeking equilibrium through ripple interactions, alignment thresholds, and consequences. What we call "past" may be meaningful to the observer, but not necessarily fundamental to the cosmos itself. The past is a way the mind arranges received consequences. It is not necessarily a physical archive existing out there as a separate region of reality.

This does not remove time from experience.

It removes time from the throne.

It asks us to stop treating time as the unquestioned ruler of cosmic interpretation.

JWST therefore becomes dangerous to the old imagination because it exposes the weakness of the time-machine metaphor. If the telescope were simply looking into a clean historical childhood, then that childhood should look like the model predicted. But it does not always do so. The deeper view shows conditions that disturb the expected sequence.

So perhaps the problem is not that the early universe was impossible.

Perhaps the problem is that "early universe" is an overloaded phrase.

It contains observation, delay, redshift, expansion, age, development, and expectation all packed into one expression. Once packed together, the phrase becomes difficult to question. But science must unpack it.

What is actually observed?

What is calculated?

What is inferred?

What is assumed?

What is narrated?

Until these are separated, the phrase "looking into the past" will continue to create artificial anomalies. We will keep finding structures that seem too mature, too bright, or too massive because we have already forced them into a developmental stage before understanding their field condition.

The telescope shows.

The model translates.

The anomaly appears in the translation.

This is the problem with calling deep space the past.

It may be partly right, but too confident. It may be useful, but too narrow. It may help beginners imagine cosmic distance, but it may mislead advanced interpretation when sharper data arrives.

JWST is asking us to become more precise.

Not poetically precise.

Scientifically precise.

It is asking us to distinguish between signal delay and cosmic history. Between redshift and absolute age. Between distance and developmental stage. Between observation and narration.

Once we do that, the collapse of the cosmic timeline begins.

Not because everything known is false.

But because the universe may be deeper than the story we built from delayed light.

2.4 — Redshift as More Than a Distance Marker

Redshift is one of the most important words in modern cosmology. It is also one of the most dangerous when interpreted too quickly.

In mainstream cosmology, redshift is commonly treated as a sign that light has been stretched as the universe expands. The more redshifted the signal, the farther the source is assumed to be. The farther the source, the older the light. The older the light, the earlier the universe. This chain has become so familiar that it often appears almost automatic.

- Redshift becomes distance.
- Distance becomes time.
- Time becomes history.
- History becomes cosmic evolution.

And then the observed object is judged according to the age assigned to it.

This is how a measurement becomes a story.

Again, the issue is not whether redshift is real. It is real as an observed shift in wavelength. The issue is not whether it carries meaningful physical information. It certainly does. The issue is whether we have reduced its meaning too narrowly by treating it mainly as a timestamp inside an expanding universe.

What if redshift is not only a sign of distance?

What if redshift also carries the signature of field condition?

What if the light we receive has not merely travelled across space, but interacted with layers of resistance, strain, density, alignment delay, and energy redistribution within the deeper cosmic field?

This is where Universal Energy Dynamics opens a different interpretation.

In UED, the universe is not treated as empty space filled with objects. It is understood as a structured energy field, a continuous medium of interaction, ripple behavior, equilibrium-seeking tendencies, and field consequences. Light is not simply a messenger flying through nothing. It is part of energy behavior moving through structured conditions.

So when light arrives redshifted, UED asks a wider question.

What has happened to the signal as a consequence of field interaction?

The mainstream reading says: the wavelength has stretched because space itself has expanded.

UED asks: could the shift also reflect field strain, ripple resistance, energy redistribution, and alignment threshold differences across regions of Structured Vacuum Energy?

This does not mean every existing redshift calculation must be casually thrown away. That would be immature. But it does mean the interpretation deserves deeper inspection, especially when the conclusions built upon it create repeated anomalies.

If redshift is treated only as a historical distance marker, then high-redshift structures must belong to an early stage of cosmic history. Once they are placed there, their brightness, mass, organization, and maturity become shocking. But if redshift also reflects field condition, then the same observation may carry a different meaning.

A high-redshift galaxy may not simply be a very ancient object seen in the childhood of the universe.

It may be a signal emerging from or passing through a high-tension field region.

Its redshift may not only tell us "how far back" we are looking. It may also tell us about the resistance profile of the field, the density of energy configuration, the intensity of ripple strain, and the delay involved in observational alignment.

This is a major shift.

Because once redshift is opened beyond distance alone, cosmic depth becomes more than chronology. It becomes structural. It becomes energetic. It becomes a measure not only of where and when, but also of what kind of field condition is involved.

Modern cosmology often treats the universe like a balloon expanding. Light stretches because the fabric stretches. The metaphor is simple and useful, but perhaps too comfortable. It gives the mind an image: galaxies moving apart, space expanding between them, wavelengths stretching during the journey.

But UED is not satisfied with the balloon metaphor.

The universe may not be an expanding fabric in the way we imagine. What we call expansion may be the large-scale observational expression of energy redistribution, vacuum gradient flow, and structural rebalancing. Redshift, in that case, may not be only the result of space stretching. It may be the visible outcome of energy interacting with the structured field through which it resolves.

This would explain why redshift cannot be treated as a dead label.

It is not merely a cosmic address.

It may be a field-condition signature.

In ordinary experience, a signal changes depending on the medium through which it travels. Sound behaves differently in air, water, and solids. Light behaves differently through glass, atmosphere, dust, plasma, and gravitational environments. We already accept that signals are not independent of conditions. Then why should cosmic signals be treated as if they travel through a neutral nothing?

The vacuum may not be nothing.

Under UED, the vacuum is structured. It is not empty absence. It is the deeper medium of field possibility. It carries the conditions through which energy becomes observable, measurable, delayed, distorted, aligned, or saturated.

So redshift must be read with humility.

It may be a compound consequence.

It may include distance. It may include expansion-like behavior. But it may also include field tension. It may include resistance. It may include alignment delay. It may include structural density. It may include the state of the medium itself.

If this is true, then many cosmic anomalies begin to look different.

A galaxy may seem too mature because we have used redshift to assign it too strict a historical position. A compact red structure may seem too bright because we interpret its signal mainly through age and distance, not through energetic field strain. A disagreement in expansion measurements may persist because the underlying field states are not uniform in the way the model assumes.

This is why redshift must be handled carefully.

It is not wrong to use redshift.

It is dangerous to imprison redshift inside one interpretation.

A measurement should open reality, not close it.

The moment redshift becomes an unquestioned timestamp, the universe is forced into a chronological courtroom. Every distant object must defend its existence against the age assigned to it. If it appears too developed, it is accused of forming too quickly. If it appears too bright, it is accused of having unusual stars. If it appears too massive, it is forced into special growth scenarios.

But perhaps the accused is innocent.

Perhaps redshift has been asked to carry more certainty than it truly provides.

This does not weaken science. It strengthens it. Science becomes more mature when it separates observation from interpretation. The observation is the redshift. The interpretation is what we say redshift means. These two must not be carelessly merged.

JWST has made this separation urgent.

Because JWST is not merely detecting redshifted signals. It is revealing structures whose assigned redshift-based historical placement creates discomfort. Instead of rushing to protect the timeline, we must ask whether the redshift itself is being read through an incomplete framework.

From the UED perspective, redshift may be one of the great clues.

Not a simple distance marker.

Not a simple age marker.

But a window into the behavior of structured energy across cosmic field layers.

Once we begin to read it that way, the universe stops looking like a broken timeline. It starts looking like a complex field of consequences, where every signal carries not only information about travel, but also information about the condition through which it has emerged.

Redshift, then, is not merely a number attached to the past.

It may be the color of resistance.

It may be the signature of field strain.

It may be the whisper of the Structured Vacuum Energy revealing how deeply reality is layered beneath our old cosmic story.

2.5 — JWST as a Field-Penetration Instrument

JWST is usually described as a telescope looking farther into space and therefore farther into time.

This description is popular, simple, and powerful. It helps people understand why the instrument matters. It tells the public that humanity has built a machine capable of seeing deeper than before, closer to the cosmic dawn, nearer to the first stars and galaxies.

But this description may be incomplete.

JWST may not be only a telescope of distance.

It may be a telescope of depth.

Not only spatial depth. Not only historical depth. But field depth.

This distinction is central to the UED reinterpretation.

In the common imagination, a telescope is like an eye pointed outward. It collects light from distant objects and gives us an image of what was once there. The better the telescope, the farther it sees. The farther it sees, the older the light. The older the light, the earlier the universe.

But JWST does something more subtle.

Its infrared sensitivity allows it to detect signals that earlier instruments could not resolve clearly. It can look through dust more effectively. It can capture highly redshifted light shifted into infrared wavelengths. It can reveal faint, compact, deep structures that remained beyond previous observational capacity.

Mainstream cosmology interprets this mainly as deeper access into the early universe.

UED asks whether this is also deeper access into the layered structure of the present universe.

This is not a decorative difference. It changes the meaning of the telescope itself.

If the universe is treated as empty space filled with objects, then JWST is simply collecting older signals from farther objects. But if the universe is understood as a Structured Vacuum Energy continuum, then observation is not merely reception of light. It is interaction with a layered field condition. The signal reaching us is not a dead postcard from a dead past. It is a consequence of energy behavior, field resistance, ripple tension, and alignment delay.

In this view, JWST becomes a field-penetration instrument.

It does not merely look backward.

It resolves deeper.

It enters layers of cosmic structure that were previously inaccessible to human observation. It reveals regions where energy conditions are more compact, more resistant, more strained, and more saturated than our local stabilized field environment. These regions appear to mainstream interpretation as ancient, early, and developmentally surprising. But UED reads them as high-tension field layers whose signals carry structural intensity.

The difference is like looking through fog.

With weak vision, the landscape appears flat and limited. With better vision, deeper forms appear. A distant mountain may suddenly emerge. The observer may feel that a new world has appeared, but the world was already there. The instrument did not create the depth. It only crossed the earlier limit of resolution.

Similarly, JWST has not created cosmic maturity at depth.

It has crossed an observational threshold.

The question is: what kind of threshold?

Mainstream cosmology says it is mainly a threshold of historical distance. UED says it is also a threshold of field resolution. JWST's infrared capability may be allowing us to perceive deeper configurations of

Structured Vacuum Energy that were not previously visible through older observational windows.

In that sense, JWST is less like an archaeological tool and more like a sensor of active structure.

Archaeology deals with remains. It studies what has been left behind. It reconstructs past civilizations from ruins, bones, tools, layers, and fragments. The cosmic archaeology metaphor has been applied widely to deep-space observation. It suggests that distant galaxies are ancient ruins preserved by light.

But the universe may not be a ruin.

It may be an active field.

JWST may not be digging up dead cosmic bones. It may be detecting active field consequences that appear distant because of observational delay but also appear strange because they belong to structural layers very different from our local field condition.

This interpretation changes the emotional tone of the discoveries.

If JWST is an archaeological tool, then mature structures at depth look like historical impossibilities. It is as if we dug into a layer expected to contain stone tools and suddenly found a modern city. The timeline collapses into confusion.

But if JWST is a field-penetration instrument, then mature structures at depth are not automatically impossible. They may represent regions where field tension, energy density, and ripple alignment generate structure differently from the slow mechanical expectation. Their brightness and compactness may be signs of intense field condition, not merely impossible chronological development.

This is why the telescope is so important for UED.

It gives observational pressure to an old assumption: that deeper means earlier in a simple linear sense.

UED does not reject depth.

It redefines what depth may include.

Depth may include signal delay, but also structural resistance.

Depth may include distance, but also field density.

Depth may include redshift, but also ripple tension.

Depth may include observational history, but also present-layer configuration.

This layered reading is more faithful to the complexity of reality. A cosmic signal does not arrive as a pure untouched message. It carries consequences of its emergence, interaction, path, medium, resistance, and resolution. To reduce all this into a simple timestamp may be convenient, but convenience is not always truth.

JWST's power lies precisely in its ability to disturb convenience.

It is not only showing us objects. It is showing us the inadequacy of one-dimensional interpretation. It is telling us that the universe cannot be read only as a timeline. It must also be read as a field of conditions.

This is why the phrase "looking back in time" must be handled carefully. It may remain useful for public explanation, but it cannot remain the final philosophical foundation of cosmology. A serious science must ask what the signal is passing through, what field state it represents, what resistance shaped it, what alignment delay affected it, and what deeper structural layer it reveals.

JWST has given us a chance to ask those questions.

It has opened a window not merely to the far universe, but to the hidden depth of the present universe. It may be showing us that reality is not arranged like a film strip, where each frame waits behind another in chronological order. It may be showing us that the cosmos is layered like a field, where deeper observation reveals different states of energy organization.

In this sense, JWST is not only an instrument of astronomy.

It is an instrument of correction.

It corrects the arrogance of assuming that the universe must appear immature at depth. It corrects the habit of turning every distant signal into a fixed historical chapter. It corrects the comfort of thinking that our timeline is the same as reality.

The telescope did not merely extend our sight.

It extended the responsibility of interpretation.

And this is where the real work begins.

Because seeing deeper is not enough.

We must also learn to think deeper.

2.6 — The Collapse of the Cosmic Movie Reel

For a long time, modern cosmology has invited us to imagine the universe as a movie reel.

The beginning is the first frame. Expansion is the motion of the film. The first particles appear, then atoms, then stars, then galaxies, then planets, then life, then observers. Every later frame is expected to be more complex than the earlier one. Every earlier frame is expected to be simpler, younger, smaller, and less mature.

This metaphor became powerful because it is easy for the human mind to understand.

We know stories. We know sequence. We know childhood, youth, maturity, and old age. We know construction, growth, development, and completion. We know that a house is not complete before its foundation, that a tree is not mature before its seed, and that a child does not become an adult in a day.

So when science arranged the universe in a similar sequence, the mind accepted it naturally.

The universe began.

The universe grew.

The universe matured.

And we are now looking back at its earlier frames.

This is the cosmic movie reel.

In this imagination, a telescope becomes a reverse projector. It does not merely look outward; it runs the film backward. The farther it sees, the earlier the frame. The earlier the frame, the more primitive the structure should be. Therefore, when JWST looks deeper, it should reveal earlier chapters of the cosmic film.

But JWST disturbed this metaphor.

It opened the supposed early frames and found structures that did not look obediently primitive. It found brightness where dimness was expected. It found compactness where roughness was expected. It found possible maturity where infancy was expected. It found density, organization, and intensity in places where the script expected slow beginnings.

This is why the discomfort became so sharp.

If the universe is truly a movie reel, then the early frames must behave like early frames. They must show the beginning of construction, not developed architecture. They must show gradual formation, not surprising maturity. They must show the first sketches, not complete patterns.

But if the early frames already contain signs of advanced structure, then something is wrong.

Either the structures formed much faster than expected, or the timeline needs major adjustment, or the interpretation of the frame itself is incomplete.

Mainstream cosmology usually begins with the first two options. It asks how galaxies formed so quickly, how stars became efficient so early, how black holes grew so rapidly, and how matter organized itself within a very short chronological window. These are valid questions inside the movie-reel model.

But UED asks the third question.

What if the movie reel itself is the wrong metaphor?

What if the universe is not fundamentally a sequence of frames moving from primitive to mature in the way the human mind imagines?

What if the universe is better understood as a structured depth-field, where different layers reveal different energy conditions, different ripple states, different resistance profiles, and different alignment thresholds?

This changes the entire reading.

In a movie reel, depth means past.

In a field, depth means condition.

In a movie reel, early means immature.

In a field, compactness may mean intensity.

In a movie reel, maturity requires duration.

In a field, structure may emerge wherever alignment reaches threshold.

In a movie reel, a distant galaxy must defend its age.

In a field, the galaxy expresses its condition.

This is the collapse of the cosmic movie reel.

It is not the collapse of observation. It is not the collapse of science. It is the collapse of one metaphor that became too powerful in our interpretation of the universe.

The universe may not be a film running from beginning to end.

It may be a continuous structured field, where what we call "past," "distance," "redshift," and "maturity" are interpretive categories applied by the observing mind.

From the UED perspective, the cosmos is not primarily a historical object. It is an active field of structured energy. It does not preserve itself as a sequence of dead scenes waiting for human telescopes to recover them. It expresses itself through field tension, energy redistribution, ripple alignment, saturation, resistance, and consequence.

In such a universe, JWST is not simply looking at earlier scenes.

It is resolving deeper layers.

Those layers may appear ancient because of signal delay. They may appear distant because of redshift. But they may also be structurally intense, field-dense, and energetically active within the continuous present. Their apparent maturity may not be an embarrassment to reality. It may be a sign that reality is not organized according to the film-strip logic we imposed on it.

The movie-reel metaphor creates artificial expectations.

It expects that if we go deep enough, we must find a simpler universe. But why must this be so? Because the model says so. Because the story says so. Because the human mind has converted cosmic depth into chronological childhood.

But the universe may not have childhood in the way we imagine.

It may have thresholds.

It may have field states.

It may have regions of high resistance and regions of low resistance.

It may have saturation zones and alignment zones.

It may have local equilibrium events and large-scale redistribution patterns.

It may not be moving through time like a character in a story. Instead, what we call time may be the mind's way of arranging consequences after energy states resolve into observable form.

This is why JWST is so disruptive.

It does not merely challenge a few calculations. It challenges the storytelling architecture beneath those calculations. It asks whether the universe is truly unfolding like a movie, or whether we have mistaken observational delay for universal narration.

A delayed signal is real.

But a delayed signal is not automatically a complete historical scene.

A redshifted object is real.

But redshift is not automatically a permission slip for every assumption we attach to it.

A distant structure is real.

But distance is not automatically immaturity.

Once these distinctions become clear, the old cosmic movie reel begins to lose its authority.

This does not mean the universe becomes meaningless. On the contrary, it becomes deeper. It becomes more alive intellectually. It can no longer be reduced to a simple sequence of old frames. It must be understood as a structured field whose signals require more than chronological decoding.

This is the movement from cosmic storytelling to cosmic interpretation.

And perhaps this is exactly where modern cosmology must evolve.

JWST has not shown us that the early universe is impossible. It has shown us that our idea of "early" may be too human, too narrative, and too dependent on the comfort of sequence.

The universe may not be saying, "I formed too fast."

It may be saying, "You are reading me as a film when I am a field."

And once that is understood, the collapse of the cosmic timeline is not a disaster.

It is liberation.

Closing Reflection — The Past May Be a Human Arrangement

Perhaps the most difficult thing for the human mind to accept is that reality may not be arranged in the same way the mind arranges experience.

We live through sequence.

Morning comes before afternoon. Childhood comes before adulthood. A seed comes before a tree. A foundation comes before a building. Memory comes before planning. Cause comes before consequence. Because of this, the mind naturally believes that reality itself must be organized as a story moving from earlier to later, from simple to complex, from beginning to maturity.

This is useful in life.

Without sequence, daily experience would become impossible to organize. Without memory, identity would collapse. Without past and future, action would lose direction. The human mind needs time as an internal arrangement. It needs history to make sense of change. It needs narration to convert scattered events into meaning.

But the universe may not be operating through human narration.

This is the deeper discomfort that JWST has brought into modern cosmology.

The telescope did not merely show distant objects. It disturbed the confidence with which distance is converted into past, past into youth, youth into immaturity, and immaturity into expectation. It forced us to ask whether the "early universe" is a direct physical reality exactly as we imagine it, or whether it is partly an arrangement created by the observing mind through its interpretation of delayed signals.

This does not mean observation is false.

It means interpretation must become more careful.

Light-travel delay is real. Redshift is real. Distant signals are real. But the full story built around them may not be as final as we assumed. The moment we say, "This galaxy belongs to the early universe," we have already done more than observe. We have placed the object inside a timeline. We have assigned it a developmental stage. We have judged its maturity according to a chronological expectation.

Then, if it does not behave as expected, we call it anomalous.

But perhaps the word "early" itself needs examination.

Early according to what?

Early according to signal arrival?

Early according to redshift calculation?

Early according to expansion interpretation?

Early according to the model's developmental sequence?

Early according to the human habit of turning distance into history?

These questions matter because JWST has exposed the risk of collapsing all these meanings into one word. When we say "early universe," the phrase sounds simple. But hidden inside it are many layers: distance, delay, redshift, expansion, age, formation history, structural development, and theoretical expectation. If even one of these layers is misunderstood, the entire interpretation becomes fragile.

Universal Energy Dynamics asks us to separate these layers.

It asks us to see that the past may not be a cosmic warehouse where old frames are preserved for our telescopes. The past may be a human arrangement of received consequences. It may be the mind's way of organizing delayed interaction into a meaningful story. In experience, this arrangement is necessary. But in cosmology, if applied too rigidly, it can become misleading.

The universe may not be storing its childhood as a visible archive.

It may be presenting layered field conditions, and the observer may be translating those conditions into history.

This is why JWST matters so much. It has not ended the search for cosmic origin. It has not made cosmology useless. It has not proved that every existing model must be thrown away. But it has weakened the casual confidence of the cosmic movie reel. It has shown that deeper observation does not always behave like earlier infancy. It has shown that the universe at cosmic depth may contain intensity, structure, compactness, and maturity that challenge the old sequence.

The past, then, may be less like a road behind us and more like a narrative arrangement created after interaction.

This is not merely a scientific issue. It is also philosophical.

The mind loves history because history gives control. If we can say where something began, how it developed, and where it is going, we feel we have understood it. But the universe may not surrender itself so easily. It may not be a storybook written for the comfort of observers. It may be a continuous field of energy consequences, where what we call before and after are ways of arranging resolution, not independent containers of reality.

Under UED, the cosmos is not dead history stretched across space. It is structured presence. It is a continuous Now, layered through field tension, ripple behavior, resistance, saturation, and alignment delay. Observation does not simply retrieve earlier frames. It participates in the unfolding of measurable consequence.

This is why JWST may not be showing us that the early universe was impossible.

It may be showing us that our idea of "early" was too human.

Too narrative.

Too dependent on sequence.

Too loyal to the comfort of a beginning-to-maturity story.

The telescope has done its work. It has looked deeper. Now interpretation must do its work. It must become deeper as well.

If we continue reading the universe only as a timeline, we will keep finding contradictions wherever field structure refuses chronological simplicity. But if we learn to read the universe as layered energy reality, many contradictions may begin to transform into clues.

The past is meaningful to the mind.

But the cosmos may not be mind's memory.

And perhaps that is the real collapse of the cosmic timeline.

Not that time disappears from experience.

Not that delay becomes meaningless.

Not that science loses direction.

But that the universe is no longer reduced to a human-style biography.

JWST has not merely opened a window into space. It has opened a crack in the way we arrange reality. Through that crack, a different possibility appears.

The universe may not be saying, "Look how early I was."

It may be saying, "Look how deeply I exist."

Chapter 3 — The Myth of the Early Galaxy

How "impossible" mature galaxies and Little Red Dots challenge the slow mechanical story of cosmic evolution.

When a Galaxy Refuses to Be Young

A galaxy does not know that we have called it early.

It does not know that our equations have placed it inside a young chapter of the universe. It does not know that our timelines expect it to be faint, small, chaotic, and structurally incomplete. It does not know that its brightness may create academic discomfort, that its compactness may trouble models, or that its maturity may disturb the neat sequence of cosmic evolution.

It simply exists.

This is where the problem begins.

Not in the galaxy.

In our expectation of the galaxy.

For a long time, modern cosmology has treated distant galaxies as younger galaxies. The deeper we look, the earlier we believe we are seeing. The earlier we believe we are seeing, the less mature the structures are expected to be. This is the logic of the cosmic timeline. It is simple, powerful, and emotionally satisfying.

But JWST has disturbed this comfort.

It has shown us deep-field structures that do not behave like obedient cosmic infants. Some appear surprisingly bright. Some appear highly compact. Some seem dense, energetic, and organized beyond what the slow mechanical story comfortably expects. Some are now discussed through terms like "Little Red Dots," as if their smallness should make them less dangerous to the timeline. But small things can carry large implications.

A tiny crack can reveal a weak wall.

A small signal can challenge a large theory.

A little red dot can become a big philosophical problem.

The issue is not merely that these objects exist. The issue is that they appear in regions where the standard interpretation expects cosmic immaturity. If the universe is a movie reel, and JWST is looking at earlier frames, then those frames should show early formation. They should show galaxies still struggling toward structure. They should show rough beginnings, not compact intensity.

But what if the "early galaxy" is already a myth created by interpretation?

What if the galaxy is not early in the way we imagine?

What if it is not a child of the universe, but a field condition being read through chronological language?

This chapter begins with that question.

A galaxy becomes "early" only after we place it inside a timeline. The object itself does not carry the label. The label comes from our reading of redshift, distance, light-travel delay, expansion, and cosmic history. Once that label is attached, the galaxy is judged by developmental expectations. If it appears mature, we call it surprising. If it appears bright, we call it unusually luminous. If it appears massive, we search for accelerated formation mechanisms.

But perhaps the first mistake is the label itself.

Under the slow mechanical story, a galaxy is imagined as something built gradually. Matter gathers. Gas collapses. Stars form. Mergers happen. Dark matter provides scaffolding. Gravity slowly organizes the architecture. Time becomes the builder. In that story, maturity must be earned through duration.

So when JWST reveals mature-looking structures at cosmic depth, the old story becomes uncomfortable.

But Universal Energy Dynamics asks us to look differently.

A galaxy may not be merely a pile of matter collected across time. It may be a large-scale field consequence. It may be ripple condensation within Structured Vacuum Energy. It may be the result of field tension reaching alignment threshold. It may be structure emerging when the deeper medium resolves disturbance into organized form.

In that case, maturity is not only a question of age.

It is a question of field condition.

This changes everything.

If structure depends only on time, then mature galaxies at high redshift are anomalies. But if structure depends on field alignment, energy density, ripple saturation, and equilibrium-seeking behavior, then such galaxies may not be anomalies at all. They may be natural expressions of intense field realignment.

The galaxy was not too early.

Our interpretation was too narrow.

The galaxy was not too mature.

Our expectation of cosmic childhood was too human.

The galaxy was not impossible.

The category we placed it in was incomplete.

This is why the myth of the early galaxy matters. It is not just a technical problem in astrophysics. It is a philosophical weakness in how we read the universe. We first decide that deep means past. Then we decide that past means young. Then we decide that young means immature. Then we observe something mature and call reality strange.

But perhaps reality is not strange.

Perhaps it is only resisting the story we built around it.

JWST has not merely shown us galaxies. It has shown us the danger of forcing galaxies to behave like children in a human biography of the universe. It has shown us that structure may not wait politely for our timelines. It has shown us that cosmic depth may carry field intensity, not just historical infancy.

A galaxy does not need our permission to be structured.

It does not need our timeline to be luminous.

It does not need our model to be real.

It simply appears as a consequence of deeper cosmic behavior.

And when such a galaxy refuses to be young, perhaps the honest response is not to accuse the galaxy of being impossible.

Perhaps the honest response is to admit that "young" was our word, not the universe's truth.

3.1 — The Slow Mechanical Story of Galaxy Formation

In the standard story of cosmology, a galaxy is not born fully formed.

It is built.

Slowly.

Step by step.

Matter gathers. Gas cools. Clouds collapse. Stars ignite. Gravity pulls. Smaller structures merge into larger structures. Dark matter provides an invisible scaffolding. Over long periods, chaotic beginnings slowly organize into recognizable galaxies. What begins as scattered material eventually becomes structure.

This is the slow mechanical story of galaxy formation.

It is a powerful story because it feels reasonable. We see similar patterns around us. A building takes time. A city takes time. A tree takes time. A civilization takes time. Complexity, in ordinary human experience, usually emerges through duration. So when cosmology says that galaxies also require long cosmic time to form, the mind easily accepts it.

The model feels natural.

First there is simplicity.

Then gradual accumulation.

Then organization.

Then maturity.

This sequence satisfies the human mind because it resembles the way we understand growth everywhere else. A child does not become an adult immediately. A seed does not become a forest overnight. A small settlement does not become a metropolis without many layers of development. So we assume the universe must also follow this kind of progressive construction.

In mainstream cosmology, gravity becomes the main organizer. Matter is distributed after the early universe cools. Slight density variations allow some regions to attract more matter than others. Gas begins to gather. Dark matter, though unseen directly, is treated as the gravitational framework within which ordinary matter can settle. Over time, the first stars form.

Those stars live, explode, enrich the surroundings, and prepare the environment for later generations of stars. Galaxies assemble through collapse, interaction, accretion, and merger.

It is a grand construction project.

And in such a project, time is the main builder.

This is why mature galaxies at great cosmic depth become a problem. If a galaxy needs long duration to become massive, bright, chemically rich, compact, or organized, then a mature-looking galaxy in the so-called early universe creates pressure. The model must explain how so much development happened so quickly.

This is not a minor issue.

The slow mechanical story does not merely describe galaxies. It depends on a particular relationship between time and structure. It assumes that structure grows from simpler states into more complex states through a long chain of events. It assumes that maturity cannot appear unless enough chronological process has taken place. It assumes that cosmic architecture must earn itself through duration.

Within this story, an early galaxy must look early.

It may be forming stars. It may be irregular. It may be smaller. It may be chemically younger. It may be unsettled. But it should not appear too mature, too bright, too dense, or too organized. If it does, then the explanation must be stretched.

Perhaps star formation was more efficient.

Perhaps gas collapsed faster.

Perhaps mergers happened earlier.

Perhaps black holes contributed to the brightness.

Perhaps dark matter halos formed in unusual ways.

Perhaps the estimates require adjustment.

These are all attempts to preserve the mechanical timeline.

Again, we should be fair. The slow mechanical story is not useless. It has explained many observations. It has given cosmology a working structure. It has allowed scientists to simulate galaxy formation, classify stages, estimate mass, model mergers, and connect small-scale processes to large-scale cosmic evolution. It is not foolish. It is not casual imagination. It is a serious scientific framework developed through enormous intellectual effort.

But every framework carries a hidden cost.

The cost of the slow mechanical story is that it becomes heavily dependent on time.

If time is the builder, then maturity must be delayed. If structure is accumulated mechanically, then complexity must come after enough accumulation. If galaxies are assembled from parts, then the earliest cosmic regions must show incomplete assembly.

This is exactly where JWST creates discomfort.

JWST shows that some deep-field structures do not politely follow the expected developmental stage. They do not always appear as unfinished sketches. Some look like concentrated signatures of strong organization. Some appear bright enough to challenge the expected star-formation history. Some appear compact enough to challenge the scale of slow assembly. Some appear massive enough to make the available timeline feel too narrow.

The model can respond by accelerating the process.

But acceleration still belongs to the same mechanical story.

It still says: the galaxy is a thing assembled from parts; it simply assembled faster than we expected.

Universal Energy Dynamics asks a deeper question.

What if a galaxy is not primarily a pile of matter slowly collected across time?

What if it is first a field consequence?

What if matter itself is not the fundamental starting point, but a structured outcome of deeper energy behavior?

What if galactic formation is not only mechanical assembly, but ripple condensation, field alignment, and equilibrium-seeking organization within Structured Vacuum Energy?

Then the problem changes completely.

In the slow mechanical story, the galaxy waits for time. In UED, structure waits for threshold.

This is a very different universe.

A threshold-based universe does not require maturity to be understood only as old age. It allows structure to emerge where field conditions are intense enough, aligned enough, saturated enough, and ready for equilibrium resolution. It does not remove process, but it changes the nature of process. The process is no longer only external accumulation. It is internal field organization.

This is why "early galaxies" become misleading.

They are called early because they are placed in a chronological frame. Once placed there, they are judged by the slow mechanical story. If they appear mature, the model becomes uncomfortable. But the discomfort may arise because the galaxy has been forced into the wrong type of explanation.

A galaxy may not be like a city slowly built brick by brick.

It may be more like a pattern emerging in a field when the conditions are right.

A whirlpool does not need bricks. It needs flow conditions.

A flame does not need construction. It needs ignition conditions.

A wave does not need assembly. It needs disturbance and medium.

Similarly, under UED, a galaxy may not require the kind of slow mechanical childhood we imagine. It may require sufficient field tension, ripple behavior, and structural alignment.

This does not mean there is no evolution within galaxies. Galaxies may still change, interact, merge, stabilize, dissolve, and transform. But their basic structural emergence may not depend only on a long chronological construction process. It may arise through deeper energy conditions that mainstream cosmology has not fully recognized.

The slow mechanical story gave us a useful map.

But JWST has shown that the map may not be enough.

If every mature-looking deep structure must be explained by special acceleration, unusual efficiency, hidden scaffolding, or exceptional conditions, then perhaps the problem is not only speed. Perhaps the problem is the belief that galaxies are mainly built by slow mechanical collection.

Maybe galaxies are not merely cosmic constructions.

Maybe they are field expressions.

And if that is true, then the "early galaxy" is not a young object that grew too fast.

It is a structured consequence that our timeline has misunderstood.

3.2 — The Shock of "Impossible" Galaxies

The word "impossible" should be used carefully in science.

Reality does not become impossible because a model struggles. The universe does not stop existing when equations become uncomfortable. A galaxy does not vanish because its brightness, mass, or structure does not fit the predicted stage of cosmic evolution.

When scientists or commentators call certain JWST galaxies "impossible," the word does not truly mean impossible for the universe.

It means difficult for the model.

This distinction is critical.

An impossible galaxy is not a cosmic crime. It is an interpretive crisis. The galaxy is not violating reality. It is violating expectation. It is not breaking the universe. It is breaking the assumed sequence through which we expected the universe to develop.

That is why the shock of these galaxies is so powerful.

They appear where the standard timeline expects infancy, but they carry signs of maturity. They appear in regions interpreted as extremely early, but they do not always behave like weak beginnings. They show brightness where faintness was expected. They show compactness where loose formation was expected. They show density where gradual assembly was expected. They show organization where cosmic roughness was expected.

This creates a very interesting philosophical situation.

The object appears.

The model measures it.

The timeline assigns its age.

Then the same timeline becomes surprised by what it has assigned.

This is like inviting a person into a room, giving him the role of a child, and then becoming shocked when he speaks like an adult. The shock does not come only from the person. It comes from the role we forced upon him.

Similarly, the galaxy becomes "impossible" only after being placed inside the expected childhood of the universe.

Before that, it is only a galaxy.

It is only light.

It is only structure.

It is only a signal.

The impossibility begins when interpretation enters.

This does not mean the observations require no careful analysis. They do. Measurements of mass, brightness, age, dust, star formation, black hole activity, and distance must be tested again and again. Science must remain disciplined. Some early claims may later be revised. Some objects may turn out to be less massive than first thought. Some may have different compositions or luminosity sources. Some may be explained within modified mainstream models.

But even after all reasonable corrections, the larger discomfort remains.

JWST has made it harder to maintain the simple idea that deep cosmic structures must be primitive.

The shock is not only in one galaxy. It is in the pattern of expectation being disturbed.

When many deep structures appear brighter, denser, or more organized than expected, the issue is no longer merely technical. It becomes structural. It asks whether our definition of "early" has become too rigid. It asks whether our assumption about the speed of structure formation is too narrow. It asks whether the universe must really follow the slow mechanical timetable we have assigned to it.

The mainstream response usually tries to make the timeline more flexible.

Perhaps galaxies formed stars faster than expected.

Perhaps early gas clouds collapsed more efficiently.

Perhaps massive stars were more common.

Perhaps black holes inside these galaxies made them appear brighter.

Perhaps dust and observational effects misled mass estimates.

Perhaps the first galaxies grew in more extreme conditions than previously imagined.

These are serious possibilities. They deserve investigation. But they still operate inside the same broad logic: the galaxy is a mechanical object that formed earlier and faster than we thought.

Universal Energy Dynamics asks a different question.

What if the galaxy is not early?

What if the word "early" is itself the beginning of the error?

What if the observed structure is not a mechanically accelerated child of the universe, but a field condition being misread as chronological infancy?

Under UED, a galaxy is not primarily understood as a collection of matter that slowly piles up across time. It is a structural consequence of energy behavior within the deeper field. Matter is not the first truth. Matter is ripple condensation. Gravity is not the only builder. Field alignment, energy gradients, ripple tension, and equilibrium-seeking behavior are part of the deeper mechanism.

In that reading, a so-called impossible galaxy does not require panic.

It requires translation.

The question changes from, "How did this galaxy grow so fast?" to "What field condition produced this structure?"

This shift is enormous.

The first question assumes the slow mechanical story is correct and searches for accelerated formation. The second question examines whether the galaxy was ever correctly understood as a slow mechanical object in the first place.

If galaxies are field consequences, then maturity does not depend only on chronological duration. It depends on the state of the field. It depends on local energy density. It depends on how ripple behavior stabilizes. It depends on how disturbance contracts toward equilibrium. It depends on whether the medium itself reaches the threshold required for structural manifestation.

This does not make galaxy formation magical.

It makes it less mechanically limited.

A pattern can emerge suddenly when conditions cross a threshold. Water can remain calm for some time, then form a whirlpool under the right flow. A cloud can appear as vapor reaches a condensation point. A flame can emerge when fuel, oxygen, and ignition meet. In such cases, structure is not built brick by brick. It appears when conditions become ready.

UED applies this broader principle at the cosmic scale.

The galaxy may be a threshold event.

Not a violation.

Not a miracle.

Not an impossible object.

A consequence.

The language of "impossible galaxies" therefore reveals the weakness of our interpretive habit. We are still judging cosmic structures as if the universe were constructing itself like a human project. We imagine raw material, gradual assembly, supporting scaffolds, and long development. Then, when structure appears too intensely or too compactly, we assume something extraordinary must have happened to speed up construction.

But perhaps the universe was never constructing in that way alone.

Perhaps it was structuring.

There is a difference.

Construction requires parts to be added from outside.

Structuring occurs when conditions inside a field reorganize into stable form.

The first depends heavily on accumulated time.

The second depends heavily on threshold and alignment.

This is why the word "impossible" may be misleading. It protects the old model by making the object look strange. But the object is not strange from its own side. It is only strange relative to the expectation imposed upon it.

A mature-looking galaxy at cosmic depth is not saying, "I formed too early."

It may be saying, "You have misunderstood formation."

A bright compact structure is not saying, "I broke your timeline."

It may be saying, "Your timeline is not the deepest mechanism."

A massive distant object is not saying, "I violated cosmic law."

It may be saying, "Your law is incomplete."

This is the real shock of the so-called impossible galaxies.

They are not shocking because they are impossible.

They are shocking because they reveal how much modern cosmology depends on a particular story of time, growth, and mechanical assembly.

If that story weakens, the galaxy becomes less impossible.

The interpretation becomes more exposed.

And once the interpretation is exposed, the next step becomes unavoidable: we must ask whether "early galaxy" is a physical truth or a label created by the timeline.

That question is where the myth begins to dissolve.

3.3 — Little Red Dots and the Problem of Compact Maturity

Sometimes the smallest phrases carry the largest disturbance.

"Little Red Dots" sounds harmless. Almost playful. Almost innocent. The phrase does not sound like a threat to cosmology. It does not sound like something that can disturb the confidence of a grand cosmic timeline. It sounds like a label one might give to tiny marks on an image, small reddish points in the deep field, interesting but manageable.

But in science, small things often become dangerous.

A small mismatch can expose a large assumption. A faint signal can change a theory. A dot can become a door. And in the case of JWST, these Little Red Dots have become more than tiny objects in distant observation. They have become concentrated symbols of a deeper problem: compact maturity where the cosmic timeline expected primitive formation.

They are little only in appearance.

They are not little in implication.

The problem begins with their character. These objects appear compact, red, bright, and energetically significant. They do not look like the weak, scattered, immature structures that the standard early-universe imagination expected to dominate cosmic depth. They appear concentrated. They appear intense. They appear as if something highly organized is present where the old story expected developmental simplicity.

This is why they matter.

If a galaxy is large and mature, one can perhaps search for a complex growth pathway. But when compact objects themselves appear too bright or too dense, the discomfort becomes sharper. The question is no longer only how big structures formed so quickly. The question becomes how such concentrated energetic behavior appears in regions assigned to early cosmic time.

The mainstream response is careful, as it should be. Scientists explore many possibilities. Perhaps these objects are dense star-forming galaxies. Perhaps they contain active galactic nuclei. Perhaps central black hole activity is contributing to their brightness. Perhaps dust is affecting their color. Perhaps unusual stellar populations are involved. Perhaps mass estimates are uncertain. Perhaps the red appearance is telling us something complex about their composition, distance, or energy source.

All these possibilities deserve investigation.

But from the perspective of this book, the deeper issue remains.

Most explanations still try to preserve the same background story. They ask: what special mechanism allowed such compact brightness to appear so early? What unusual condition allowed the timeline to survive? What hidden process helped these objects grow, shine, or concentrate faster than expected?

Again, the telescope observes.

The model assigns them to the early universe.

Then the model becomes surprised by their maturity.

This is the repeated pattern.

Little Red Dots are important because they compress the problem into a visible symbol. A tiny red object becomes a challenge to a massive intellectual structure. Its compactness refuses the expectation of cosmic looseness. Its brightness refuses the expectation of faintness. Its possible density refuses the expectation of slow assembly.

In that sense, they are not merely objects.

They are questions.

They ask whether cosmic structure always requires the long mechanical route imagined by the standard model. They ask whether compact energetic systems must be explained only through accelerated formation. They ask whether the early universe is truly early in the way we have been trained to think. They ask whether redshifted depth is being over-translated into chronological infancy.

Under Universal Energy Dynamics, Little Red Dots are not treated as embarrassing exceptions. They are treated as signals of field intensity.

A compact object does not necessarily mean a small mechanical object that somehow grew too fast. It may represent a localized field condition where ripple tension, energy density, and structural alignment have reached a high threshold. The redness may not only speak of distance and time. It may also speak of field resistance, strain, and the condition through which the signal becomes observable.

In UED, structure is not merely built by collecting matter. Structure emerges when the field reaches the right condition for manifestation. Matter itself is ripple condensation within Structured Vacuum Energy. Brightness is not merely the output of old stars or accelerated formation. It may also be energetic feedback from concentrated field realignment. Compactness is not merely a size problem. It may be a sign of localized saturation.

This makes Little Red Dots conceptually powerful.

They may be small regions where the deeper field has condensed into intense structural expression. They may be high-density consequences of SVE behavior. They may be points where the universe is not showing us premature galaxies, but revealing compact alignment zones within the continuous present field.

The old model asks: how did so much happen so early?

UED asks: what kind of field condition produces this compact consequence?

The old model sees them as strange because it expects early cosmic objects to be primitive.

UED sees them as meaningful because intense field conditions need not wait for our chronological comfort.

This difference matters.

If Little Red Dots are interpreted only through the slow mechanical story, then they become puzzles requiring special rescue. One explanation may say they are powered by hidden black holes. Another may say they are extreme star factories. Another may revise their masses. Another may adjust their developmental pathway. All these discussions are useful within the existing framework, but they may still avoid the deepest question.

Why did compact maturity become surprising in the first place?

The answer is simple: because we assumed that depth means early, and early means immature.

Once that assumption is loosened, the objects look different.

They no longer appear as rebellious children of the early universe. They appear as compact expressions of field behavior. They become examples of how structure can arise where conditions are ready, not merely where time has passed. They suggest that maturity may not always be a long chronological achievement. It may sometimes be a threshold consequence.

This does not remove mystery.

It removes unnecessary contradiction.

Little Red Dots may still require detailed study. Their exact nature may vary from object to object. Some may be dominated by star formation. Some may involve central regulators. Some may involve dust, density, or complex energetic processes. UED does not need to flatten all of them into one simplistic explanation.

But UED changes the foundation of the question.

Instead of forcing each dot to defend itself against the cosmic timeline, UED asks what the dot reveals about the field.

That is a more open scientific posture.

It allows the object to speak before the model sentences it.

And perhaps that is exactly what JWST is teaching us. It is showing that the universe is not afraid of compactness, brightness, or maturity at depth. These are not crimes against cosmology. They are signals of deeper structure.

The phrase Little Red Dots may sound modest.

But beneath that modest phrase lies a powerful challenge.

If such compact energetic objects can appear where the timeline expected infancy, then perhaps the issue is not their existence. Perhaps the issue is our habit of treating cosmic depth as a nursery.

The dots are little.

The assumption they disturb is enormous.

3.4 — Matter as Ripple Condensation, Not Collected Dust

The entire problem of the "early galaxy" begins with what we think matter is.

If matter is imagined as independent physical stuff scattered in space, then galaxy formation naturally becomes a collection story. Matter must gather. Gas must collapse. Stars must form. Structures must merge. Mass must accumulate. Gravity must pull everything together over a long duration. In this view, a galaxy is like a cosmic city built from available material.

First there is dust.

Then there are clouds.

Then stars.

Then clusters.

Then larger structures.

Then galaxies.

The sequence looks logical because it follows the construction model of reality. Things are built by adding parts. Bigger things require more parts. More parts require more gathering. Gathering requires time. Therefore, a mature galaxy must require long time.

This is the hidden foundation of the slow mechanical story.

But Universal Energy Dynamics begins from a different point.

In UED, matter is not treated as the first truth of reality. Matter is a consequence. It is a structured manifestation of deeper energy behavior. It is not dead substance sitting independently inside empty space. It is ripple condensation within Structured Vacuum Energy.

This changes the entire meaning of galaxy formation.

If matter itself is a field consequence, then a galaxy is not merely a collection of matter. It is a large-scale condensation pattern inside a structured field. It is a region where energy behavior, ripple interaction, tension, resistance, and equilibrium-seeking dynamics have organized into observable form.

The question then changes.

We no longer ask only, "How much matter collected over how much time?"

We ask, "What field condition produced this matter-structure consequence?"

This is a deeper question.

It does not deny that matter appears. It does not deny that stars exist. It does not deny that galaxies have measurable mass, luminosity, motion, and structure. But it refuses to treat those measurable features as final explanations. It asks what stands beneath them.

In the ordinary model, matter is the building block.

In UED, matter is the visible footprint of field organization.

This is why mature-looking deep galaxies become less shocking under UED. If galaxies are built only through external collection, then time becomes essential. But if galaxies emerge through field condensation and alignment, then the key factor is not merely duration. The key factor is threshold.

A field does not need to wait for billions of years simply because human imagination expects slow assembly. A field responds to condition. When disturbance reaches a certain intensity, when energy gradients become sufficiently active, when ripple tension crosses the necessary boundary, and when alignment becomes possible, structure can appear as consequence.

This does not mean structure appears magically.

It means the process is not the same as mechanical collection.

A cloud does not form because someone collects droplets one by one and places them in the sky. It forms when conditions allow vapor to condense. A crystal does not grow by random decoration alone. It follows internal order when the surrounding conditions permit crystallization. A whirlpool does not need a blueprint drawn from outside. It appears when flow, pressure, boundary, and motion create the right condition.

In each case, structure appears because the medium itself participates.

UED applies this principle at a cosmic scale.

Structured Vacuum Energy is not empty nothingness. It is the deeper field-medium within which energy behavior becomes manifest. When this medium experiences tension, disturbance, gradient, or imbalance, it does not remain passive. It seeks balance. It responds through ripple interaction.

It contracts delta. It moves toward equilibrium. Matter appears as condensed structure within this process.

So a galaxy is not only something made of matter.

It is a field event stabilized as matter.

This is where the phrase "ripple condensation" becomes important.

A ripple is not a solid object, yet it is real as a disturbance pattern. When ripple behavior becomes structured, repeated, stabilized, and held within field conditions, it can manifest as measurable form. What we call matter is therefore not separate from energy dynamics. It is energy held in a condensed, structured state.

From this perspective, the mature galaxy is not a pile of dust that somehow gathered too fast. It is a field condensation that reached structural expression under intense conditions.

Now the "early galaxy" problem begins to dissolve.

The old model says: there was not enough time for so much matter to assemble.

UED asks: was assembly the correct starting assumption?

The old model says: this mass should not exist so soon.

UED asks: is mass being understood as collected stuff, or as stabilized field consequence?

The old model says: this galaxy looks too mature.

UED asks: is maturity a measure of age, or a measure of achieved structural alignment?

This distinction is crucial.

Under a collection-based model, maturity is delayed by necessity. The galaxy must wait for material to gather, processes to occur, stars to cycle, mergers to happen, and structure to settle. But under a field-condensation model, maturity can arise when field conditions become sufficiently organized. It is not independent of process, but the process is internal to the field, not merely external accumulation.

This also changes how we understand brightness.

A highly luminous deep object may not only indicate many old stars or extreme star formation. It may indicate energetic feedback from concentrated field realignment. When a high-tension field region moves toward structure, energy expression can be intense. The brightness is not necessarily a sign of chronological age. It may be a sign of active condensation, resistance release, and ripple stabilization.

Similarly, compactness is not necessarily a sign of impossible compression through rapid mechanical growth. It may be a sign that the field has concentrated into a high-density alignment zone. In such a region, matter appears not because scattered material slowly wandered into place, but because the field condition itself supports condensation.

This is why Little Red Dots are so important.

They are compact. They are intense. They are red. They are difficult to fit neatly into the slow mechanical story. But under UED, they can be read as localized condensation zones inside Structured Vacuum Energy. Their small appearance does not make them weak. Their compactness may actually reveal high field tension. Their luminosity may reflect energetic release during structural alignment. Their redness may be linked not only to distance, but also to resistance and field condition.

The deeper point is this: matter cannot be separated from the field that manifests it.

Mainstream cosmology often treats space as a stage and matter as the actor. UED does not accept this separation. The stage itself is not empty. The field is active. Matter is not merely moving through the field; matter is a consequence of field dynamics. The actor and the stage are not fundamentally separate.

This has huge implications for galaxy formation.

If matter is field condensation, then galaxy formation is not simply material aggregation. It is field organization becoming visible. The galaxy is not assembled from outside like a machine. It is expressed from within the field like a pattern.

This is why the phrase "collected dust" is too weak.

Of course dust exists. Gas exists. Stars exist. Material processes exist. But they are surface-level expressions of deeper energy dynamics. To explain a galaxy only by the collection of material parts is like explaining a song only by the vibration of air molecules. Technically, something is correct, but the deeper organizing pattern is missing.

A galaxy is not merely accumulated matter.

It is structured consequence.

It is the visible geometry of field behavior.

Once this is understood, the myth of the early galaxy weakens further. The galaxy no longer has to be judged only by how long matter had to collect. It can be studied by asking what kind of field condition made such condensation possible.

This is not a small correction.

It is a complete change in the starting point.

The universe may not build galaxies like humans build walls.

It may manifest galaxies like fields resolve tension.

And if that is true, then JWST has not found galaxies that grew too quickly.

It has found field consequences that our matter-first imagination was too slow to understand.

3.5 — Field Realignment and Instant Structural Blueprint

The word "instant" is dangerous.

If misunderstood, it may sound like magic. It may sound as if a galaxy suddenly appears from nowhere without process, without mechanism, without condition, without discipline. That is not the meaning here.

In Universal Energy Dynamics, "instant" does not mean careless appearance.

It means threshold-based emergence.

A structure may appear suddenly to observation when the deeper field conditions required for that structure have already reached alignment. The visible formation may look rapid, but the rapidity belongs to the moment of manifestation, not to the absence of process. The process is not always slow external accumulation. Sometimes the process is internal field preparation.

This is difficult for the mechanical mind to accept because we are trained to imagine structure as construction. A building is built brick by brick. A road is laid layer by layer. A machine is assembled part by part. From this experience, we assume that all large structures must follow the same logic.

But nature often behaves differently.

A cloud forms when atmospheric conditions reach condensation. A flame appears when fuel, oxygen, and ignition cross threshold. A whirlpool emerges when flow and resistance arrange themselves into a rotating pattern. A crystal forms when molecular conditions allow order to express itself. In each case, the structure is not assembled from outside like a machine. It emerges when the medium itself becomes ready.

The structure was not created by magic.

It was released by condition.

This is the key to understanding field realignment.

Under UED, the universe is not an empty stage where matter pieces slowly gather like scattered bricks. The universe is a structured field. It contains gradients, tensions, resistances, ripple behaviors, saturation points, and equilibrium-seeking tendencies. When a region of Structured Vacuum Energy carries intense disturbance or imbalance, it does not remain passive. It seeks resolution. It contracts delta. It moves toward balance.

This movement toward balance is not random. It follows the internal logic of the field.

The field carries possibility before matter appears. The structure is not drawn from outside. It is latent within the condition of the field itself. When the required threshold is reached, ripple behavior begins to align. Condensation begins to stabilize. Energy gradients begin to organize. A measurable structure appears.

This is why a galaxy may not need to wait for the slow childhood imagined by mechanical cosmology.

It may wait for threshold.

The slow mechanical story says a galaxy must be gradually built from smaller parts. UED says a galaxy may emerge as a large-scale structural blueprint when the field reaches sufficient alignment. This does not deny later evolution, interaction, star formation, or internal transformation. A galaxy can still develop after manifestation. But its foundational structure may arise through field realignment rather than simple material assembly.

The blueprint is not a plan written on paper.

It is the geometry of energy behavior.

Every field has a pattern of possible resolution. When tension rises, energy does not dissolve into meaningless chaos. It seeks the most stable available arrangement. The galaxy, under UED, is one such arrangement. It is a stabilized expression of field dynamics. Its arms, density, luminosity, rotation, central regulation, and distribution are not merely accidental outcomes of collected dust. They are visible signatures of deeper energy geometry.

This is where Chavan's Limit becomes relevant.

A structure can manifest only when conditions cross the lower threshold of manifestation. Below that threshold, the field may remain unexpressed, stable, or inactive from the standpoint of observation. Once the threshold is crossed, the field begins to express form. If the energy condition becomes too intense, unstable, or saturated, the structure may approach a higher threshold where coherence collapses or transforms.

Between these limits, structure exists.

A galaxy, therefore, can be understood as a field structure operating within manifestation limits. It appears when energy conditions allow ripple condensation and alignment. It sustains as long as the field configuration remains coherent. It changes as gradients, interactions, and equilibrium pressures change.

This is very different from saying that a galaxy is simply built over time.

Time participates, but it is not the sole builder.

The deeper builder is alignment.

This helps explain why JWST's deep-field galaxies and Little Red Dots appear so disruptive. If we assume every galaxy must be mechanically assembled through long chronological accumulation, then compact maturity at high redshift becomes a crisis. But if field realignment can establish structural blueprints rapidly when thresholds are crossed, then such compact energetic objects are no longer impossible. They become expected signs of intense field behavior.

The "instant" quality is therefore not magical speed.

It is threshold response.

A field can remain in one state for a long duration, then reorganize rapidly when conditions cross the boundary. Human life also gives examples of this. A person may think silently for years, then one moment of clarity changes everything. A society may carry pressure for decades, then one event triggers reform or revolution. Water may heat gradually, then suddenly boil. The visible transition looks sudden, but the condition was building underneath.

The same logic can apply cosmically.

A region of Structured Vacuum Energy may carry tension, resistance, and ripple instability. As the field seeks equilibrium, it may reach a point where scattered behavior becomes aligned behavior. Once alignment begins, structure can appear with surprising intensity. This is not because the universe skipped process. It is because the process was not the slow mechanical process we expected.

The old model asks: how did the galaxy assemble so quickly?

UED asks: what threshold allowed the field to express structure?

The old model asks: where did enough matter come from?

UED asks: how did energy condense into matter-form under that field condition?

The old model asks: why does it look mature so early?

UED asks: why are we measuring maturity only through age and not through alignment?

This is the essential shift.

Maturity in UED does not mean oldness alone. It means structural coherence. A young-looking object may be incoherent. An old-looking object may be unstable. A compact field structure may be highly coherent despite being assigned an early chronological position. So when JWST reveals galaxies that seem mature at cosmic depth, UED does not first accuse them of violating time. It reads them as possible expressions of achieved field coherence.

This is why the phrase "instant structural blueprint" matters.

A blueprint does not mean the entire final complexity is finished forever. It means the organizing geometry has appeared. Once the blueprint is established, further processes can unfold within it. Stars may form, redistribute, collapse, stabilize, and transform. Central regulatory zones may develop or become visible. Ripple behavior may continue adjusting. The galaxy may evolve.

But the basic field identity is already there.

This explains why some deep structures may appear more organized than expected. Their organization may not be the final product of long mechanical history. It may be the initial expression of field alignment. The universe may not first create chaotic material and then slowly teach it structure. The field may carry structural tendency from the beginning of manifestation itself.

This is where mainstream cosmology often underestimates the medium.

It treats the vacuum too passively. It treats matter too independently. It treats time too dominantly. It treats gravity as the main organizer after matter already exists. UED reverses the depth of explanation. The field is active. Matter is a consequence. Gravity is gradient behavior. Time is alignment delay. Structure is field resolution.

Once this is accepted, the "early galaxy" becomes less mysterious.

It is not a fully grown city appearing in a prehistoric layer.

It is a field blueprint becoming observable at a deeper resolution layer.

And the Little Red Dots become especially meaningful. Their compactness may indicate high field concentration. Their brightness may indicate energetic release from structural realignment. Their redness may carry more than distance; it may carry resistance. They may be small in image, but large in principle because they show that structural manifestation can be concentrated, intense, and threshold-driven.

The galaxy does not need billions of years merely to earn permission to exist.

It needs the right field condition.

This is the heart of the UED reinterpretation.

The universe does not always build slowly like a mason.

Sometimes it aligns like a field.

And when alignment crosses threshold, structure appears with an authority that looks impossible only to those who expected bricks.

3.6 — Why "Early" May Be the Wrong Word

The word "early" looks harmless.

It appears simple, almost innocent. Early means before. Early means beginning. Early means young. Early means not yet mature. When we say "early galaxy," the mind immediately forms an image: a young structure, a cosmic infant, a first attempt by the universe to organize matter into form.

But this one word carries an entire universe of assumptions.

"Early" is not merely a description. It is a placement. It places the galaxy inside a timeline before we fully understand what the observation may actually represent. It does not simply say, "This signal has reached us after a long delay." It quietly says much more. It says the galaxy belongs to an earlier stage of universal development. It says the structure should be younger. It says maturity should be limited. It says complexity should still be under construction.

This is why the word becomes dangerous.

Because once a galaxy is called early, it is no longer allowed to simply be what it is. It must behave according to the expectations attached to earlyness.

If it is bright, we ask why it is so bright so soon.

If it is compact, we ask how it became so dense so early.

If it appears massive, we ask how enough matter gathered in such limited time.

If it appears organized, we ask how structure formed before the universe had enough opportunity to build it.

In every question, the word "early" silently controls the interpretation.

But what if the word itself is the trap?

What if high-redshift galaxies are not "early" in the simple developmental sense we have assumed? What if they are not immature historical objects from the childhood of the cosmos, but deep field conditions observed through delay, redshift, resistance, and alignment threshold? What if JWST is not only seeing back into a young universe, but resolving deeper layers of the present Structured Vacuum Energy field?

Then "early galaxy" becomes an incomplete label.

It may still describe part of the observational relation, but it does not define the full physical reality of the object. It tells us how the signal is arranged within our model, but it may not tell us the complete field condition behind the signal.

This distinction matters.

A label can guide thought, but it can also imprison it.

When we call something an early galaxy, we stop seeing it as a field expression and start seeing it as a chronological child. Then any maturity becomes suspicious. Any compactness becomes difficult. Any brightness becomes excessive. Any structural coherence becomes a problem.

But the galaxy itself is not early in its own experience.

It is not waiting inside a historical museum. It is not preserved as a dead cosmic photograph for human telescopes. It is not holding a certificate saying, "I belong to the primitive universe." These are interpretive categories added by the observing mind.

Universal Energy Dynamics asks us to become more precise.

From the UED perspective, the universe is not primarily a chronological movie reel. It is a structured field of energy consequences. Observation does not merely retrieve old scenes; it interacts with signals emerging through layered field conditions. Redshift does not only mark distance; it may also carry strain, resistance, and ripple-state information. Time is not an absolute road; it is alignment delay, the measurable resolution gap between energy states and observed consequence.

In such a universe, the word "early" loses its absolute authority.

A deep object may be distant in observational relation. Its signal may involve delay. Its redshift may be high. But its apparent maturity may not need to be forced into the emotional logic of cosmic youth. It may represent a field region where structure has already achieved alignment because the conditions demanded it.

This is the heart of the problem.

Mainstream cosmology often treats "early" as if it automatically means "undeveloped." But early is a temporal word. Development is a structural condition. The two are related only if the model linking them is correct. If the model is incomplete, then the relationship becomes unreliable.

A classroom example may help.

Suppose a teacher expects all students of a certain age to solve only simple problems. Then one student of that age solves an advanced problem. The teacher may say, "This is impossible for your age." But the student is not impossible. The teacher's developmental expectation was incomplete. Age was used as a proxy for capacity, but capacity did not obey the proxy.

In the same way, cosmology uses redshift and distance as proxies for cosmic developmental stage. But if mature-looking structures appear at those depths, perhaps the proxy is incomplete.

The galaxy is not impossible.

The proxy is overconfident.

This is why the myth of the early galaxy is so important. It reveals how easily observation becomes narration. We do not merely observe a galaxy. We assign it a role. We make it a child of the early universe. Then we become disturbed when it refuses to act like a child.

But perhaps it was never a child.

Perhaps it was a high-tension field structure.

Perhaps it was a localized ripple condensation.

Perhaps it was a region of Structured Vacuum Energy achieving equilibrium through compact manifestation.

Perhaps its brightness, redness, density, and coherence are not signs of impossible age, but signs of field condition.

Under UED, the universe does not need to wait for chronological permission before forming structure. Structure emerges when field conditions cross the threshold required for manifestation. If energy gradients, ripple tension, resistance, and alignment conditions are ready, then form can appear. The timing we assign afterward may not be the deepest cause.

This does not mean all conventional age estimates become meaningless. It means they must be handled with humility. They are not reality itself. They are model-based readings. They are interpretations of signals through assumptions about expansion, redshift, distance, and time.

The danger begins when interpretation becomes identity.

A high-redshift galaxy becomes "early."

An intense object becomes "too bright."

A compact structure becomes "too mature."

A model-dependent placement becomes a physical judgment.

Then the universe appears full of anomalies.

But if we remove the unnecessary confidence from the word early, many anomalies soften. The galaxy is no longer accused of being ahead of schedule. It is examined as a field consequence. Its structure is not measured only against age, but against energy condition. Its maturity is not treated as a violation, but as information.

This is a more honest science.

It does not force the object to defend itself against our timeline. It allows the object to reveal the limits of the timeline.

And this is where JWST becomes revolutionary. It has not simply found galaxies from the early universe. It has exposed the weakness of the word early. It has shown that cosmic depth may not be safely reduced to developmental childhood. It has forced us to ask whether our labels are smaller than the reality they claim to describe.

The word early belongs to the observer's arrangement.

The galaxy belongs to the universe.

Those two are not always the same.

So perhaps the problem was never early galaxies.

Perhaps the problem was the confidence with which we called them early.

Perhaps the universe is not showing us young galaxies that grew too fast.

Perhaps it is showing us structured field consequences that our timeline named too quickly.

And if that is true, then the myth begins to dissolve.

The galaxy was not early.

The interpretation was premature.

Closing Reflection — The Galaxy Was Never Late

A galaxy cannot be late.

It cannot arrive before schedule or after schedule. It cannot violate a timetable it never accepted. It cannot embarrass a theory, challenge a model, or disturb a timeline by intention. It has no obligation to appear young because we call it early. It has no duty to look immature because our equations place it near the beginning of the cosmic story.

It simply exists as consequence.

The confusion begins when we ask the galaxy to behave according to our expectation of its age.

This is the quiet mistake hidden inside the phrase "early galaxy." We do not merely observe a structure. We first place it inside a cosmic timeline. Then we assign it a developmental role. Then we compare its brightness, compactness, mass, and organization with what that role allows. If it does not obey, we call it surprising. If it strongly disobeys, we call it impossible.

But perhaps the galaxy was never impossible.

Perhaps our category was too small.

The word early carries the emotional weight of childhood. When we hear it, we imagine incompleteness. We expect roughness. We expect weakness. We expect something still becoming itself. That expectation works in many human situations because biological and social development often require sequence. A child becomes an adult over time. A village becomes a city over time. A machine is built part by part. A book is written page by page.

But the universe may not build all structures like a human project.

A field does not always behave like a construction site.

A field responds to condition.

When the condition is ready, structure appears. When tension reaches threshold, alignment begins. When disturbance seeks resolution, pattern emerges. When energy condenses within a structured medium, matter becomes measurable. When ripple behavior stabilizes, form becomes visible.

This is the UED shift.

The galaxy is not merely a pile of matter that gathered too quickly. It may be a field consequence that became visible when Structured Vacuum Energy reached the necessary threshold for ripple condensation and alignment. Its apparent maturity may not be a crime against time. It may be evidence of achieved coherence.

This is why JWST is so important.

It did not merely show us distant galaxies. It forced us to question the labels we placed upon them. It showed that some structures at cosmic depth do not fit the emotional simplicity of cosmic childhood. It showed brightness where the timeline expected faintness. It showed compactness where the timeline expected loose beginning. It showed possible structural maturity where the timeline expected gradual assembly.

The old interpretation asks, "How did the galaxy become mature so early?"

A deeper interpretation asks, "Was early the right word?"

That question changes everything.

Because once the word early is weakened, the galaxy is released from a false trial. It no longer needs to defend its existence against our schedule. It can be studied as a field structure, as a consequence of energy behavior, as a signature of alignment, as a point where the universe is revealing not merely history, but depth.

The Little Red Dots strengthen this lesson.

Their small appearance should not mislead us. They carry a large challenge. They remind us that compactness can be powerful, that brightness can emerge from concentrated field conditions, and that cosmic depth may carry intensity rather than infancy. They expose the weakness of assuming that the deepest observations must always show primitive stages.

Perhaps the universe is not showing us young galaxies that grew too fast.

Perhaps it is showing us field structures that our timeline named too early.

This does not remove all mystery. It should not. A serious framework must leave room for further observation, correction, and refinement. JWST data must still be examined carefully. Measurements must be checked. Luminosity sources must be understood. Redshift interpretations must be studied. The nature of compact objects must be tested.

But the philosophical foundation must also be tested.

Science cannot only refine numbers while protecting assumptions.

If the same type of discomfort keeps returning, then the model's language deserves examination. And in this case, the language of "early," "mature," "too massive," and "impossible" reveals that cosmology may still be trapped inside a human developmental metaphor.

The universe may be deeper than that.

It may not begin with matter as dead dust waiting to gather. It may begin with structured energy. It may not depend only on time as the builder. It may depend on thresholds of alignment. It may not produce galaxies as slow mechanical accidents. It may manifest them as field consequences within a larger equilibrium-seeking reality.

In that view, the galaxy was never late.

It was never premature.

It was never guilty of growing too quickly.

It was only misread through a timeline that arrived before understanding.

The myth of the early galaxy dissolves when we stop asking the universe to behave like a child in our story. A galaxy is not young because we call it young. It is not mature because we feel it should not be mature. It is not impossible because our model becomes uncomfortable.

It is real.

And reality must be the starting point.

Not the label.

Not the expectation.

Not the timeline.

The galaxy was never late.

Our interpretation arrived too slowly.

Interlude

Before the Lens Changes

There is a moment in every serious inquiry when questioning is no longer enough.

At first, we question because something feels wrong. Then we question because the explanation seems stretched. Then we question because the same discomfort returns again and again from different directions. But after a point, questioning alone becomes incomplete. If the old interpretation is under pressure, a new interpretation must be prepared with discipline.

This part of this book did not try to destroy cosmology.

It tried to expose a habit.

The habit is simple: whenever we look deeper into space, we immediately convert depth into past, past into youth, youth into immaturity, and immaturity into expectation. Then, when the universe refuses that expectation, we call it anomalous.

But perhaps the anomaly was never in the universe.

Perhaps it was in the conversion.

The first three chapters asked us to pause before accepting words like early, impossible, too massive, too bright, too mature, and too soon. These words sound scientific, but they often carry hidden assumptions. A galaxy becomes too mature only after we place it inside a timeline. A structure becomes impossible only after the model decides what should be possible. A Little Red Dot becomes disturbing only because we expected cosmic depth to behave like cosmic infancy.

This is not a small issue.

It touches the foundation of interpretation itself.

If JWST had merely shown one strange object, the old model could absorb it. If one galaxy looked unusual, we could call it an exception. If one estimate looked extreme, we could revise it. But when deeper observation repeatedly disturbs the developmental expectations of the timeline, intelligence must ask whether the timeline is reading reality correctly.

That is where second part begins.

Universal Energy Dynamics does not enter as a patch.

It does not enter to decorate the old story with new terminology. It does not say, "Keep everything as it is, but add one more invisible helper." That is not enough. The issue is deeper. The issue is the starting point.

What is matter?

What is gravity?

What is vacuum?

What is time?

What is structure?

What is redshift really telling us?

What do we mean when we call something distant, early, mature, or impossible?

The old interpretation begins with objects, particles, spacetime, expansion, and chronology. It imagines the universe as a construction project moving through time. Matter gathers, gravity pulls, stars form, galaxies assemble, and complexity slowly grows.

UED begins from a different place.

It begins with structured energy.

It begins with the idea that the vacuum is not empty absence, but an active field condition. It begins with the understanding that matter is not the first reality, but a consequence of energy condensation. It treats gravity not as a mysterious pull or only a bending of spacetime, but as an emergent effect of energy gradients seeking balance. It sees time not as a cosmic road, but as alignment delay. It sees structure not merely as assembled matter, but as field resolution.

Once the lens changes, the same universe begins to speak differently.

A galaxy is no longer merely a pile of collected matter.

A black hole is no longer automatically a singularity.

Redshift is no longer only a timestamp.

A mature structure at cosmic depth is no longer automatically impossible.

The universe stops looking like a broken timeline and begins to look like a layered field of consequences.

This is the shift required now.

This part loosened the old frame.

Next part will introduce the new lens.

The reader should not rush here. This transition matters. It is easy to criticize an old model. It is much harder to offer a new grammar that can carry observation, meaning, and mechanism together. A serious reinterpretation must not be built on emotion. It must be built on conceptual clarity.

So before we move forward, one thing must be understood clearly.

UED is not asking us to reject observation.

It is asking us to reinterpret what observation means.

JWST still sees what it sees. Light still arrives. Redshift is still measured. Structures still appear. Galaxies still shine. Cosmic signals still demand explanation. But the question is no longer whether these observations exist. The question is whether the old language is capable of understanding them fully.

Perhaps it is not.

Perhaps this is why the anomalies keep returning.

Perhaps the universe is not asking for another adjustment inside the same script.

Perhaps it is asking us to change the script.

Next part begins with matter because that is where the misunderstanding becomes visible first. If matter is treated as dead substance collected across time, galaxies must be built slowly. But if matter is understood as ripple condensation within Structured Vacuum Energy, then the entire story of structure changes. Galaxy formation, gravity, black holes, redshift, and cosmic maturity begin to open into a different logic.

This does not end mystery.

It gives mystery a better language.

The old lens gave us a universe of delayed light, distant history, slow assembly, hidden scaffolds, and difficult anomalies.

The new lens asks us to see a universe of field tension, ripple alignment, energy gradients, saturation zones, and equilibrium-seeking consequences.

The universe has not changed.

The telescope has not lied.

Only the lens is about to shift.

And once the lens changes, the word anomaly may begin to lose its power.

Part II — The UED Reinterpretation

"When the field is understood before the object, the anomaly begins to lose its disguise."

Chapter 4 — Matter Is Not a Thing, It Is a Field Consequence

Matter as ripple condensation within Structured Vacuum Energy, not an independent substance carrying fixed intrinsic identity.

The Mistake of Treating Matter as the Beginning

Matter is the most convincing illusion of reality.

Not illusion in the sense that it does not exist. Matter exists. A stone is felt. A body is touched. A planet is measured. A star shines. A galaxy appears across the darkness with structure, luminosity, and motion. Matter is not imaginary.

But the illusion begins when we treat matter as the beginning.

Because matter is visible, we assume it is primary. Because it can be measured, we assume it is fundamental. Because it appears stable, we assume it carries independent identity. Because our senses meet matter first, our thinking also begins with matter first.

This is natural.

A child does not begin by understanding fields. A child begins by touching objects. A cup, a table, a toy, a stone, a wall, a hand, a face. The world first arrives as things. Later, the mind learns relations, forces, motion, patterns, and systems. But the earliest impression remains powerful: reality appears to be made of objects.

Science refined this habit but did not completely escape it.

The ancient mind saw matter as elements. The classical mind saw matter as bodies. The modern mind saw matter as particles. Even when physics became more advanced and began speaking of fields, quanta, uncertainty, and interactions, the imagination of the public—and often the hidden intuition of science—continued to treat matter as the main actor.

Matter exists.

Then matter moves.

Then matter attracts.

Then matter collides.

Then matter gathers.

Then matter forms stars.

Then matter builds galaxies.

This is the matter-first universe.

It appears logical because it begins with what is visible. But visibility is not the same as fundamentality. The most visible layer of reality may not be the deepest layer. The wave seen on the ocean is real, but it is not separate from the water. The flame is visible, but it is not independent of conditions. The cloud has shape, but it is not a permanent object sitting independently in the sky. It is a condensation event.

Matter may also be like that.

Universal Energy Dynamics begins from this reversal.

Matter is not the first reality. Matter is a consequence. It is not an independent substance carrying fixed identity by itself. It is structured energy appearing as form. It is ripple condensation within Structured Vacuum Energy. It is the visible outcome of deeper field behavior.

This one shift changes everything.

If matter is primary, then the universe must be explained as the movement and gathering of matter. Galaxies become collections. Stars become accumulations. Gravity becomes the organizer of mass. Time becomes the builder of structure. Cosmic evolution becomes a slow mechanical construction project.

But if matter is consequence, the question changes.

We no longer ask only how matter gathered.

We ask what field condition produced matter.

We no longer ask only how mass accumulated.

We ask how energy stabilized into measurable form.

We no longer ask only how galaxies were built from material.

We ask how Structured Vacuum Energy condensed into large-scale patterns of structure.

This is why the issue matters so deeply for JWST and the so-called early galaxies.

If galaxies are understood only as collected matter, then mature structures at cosmic depth become difficult. There seems to be too little time for enough matter to gather, cool, ignite, merge, and organize. The model becomes forced to accelerate the process. It must imagine faster star formation, heavier seeds, unusual efficiencies, hidden scaffolds, or exceptional conditions.

But if matter itself is ripple condensation within a structured field, then the galaxy is not merely a collection of things. It is a field consequence. Its maturity may not depend only on how long material had to gather. It may depend on how field tension, energy density, ripple behavior, and alignment conditions reached structural threshold.

The universe may not be building matter like bricks.

It may be expressing matter like patterns.

This does not make matter less real. In fact, it makes matter more meaningful. Matter becomes the visible language of deeper energy behavior. A star is not just a hot object. A planet is not just a sphere of material. A galaxy is not just a crowd of stars. Each is a stabilized expression of field dynamics.

The mistake of treating matter as the beginning has shaped our interpretation of the cosmos for too long.

It has made us imagine the universe as a warehouse of things moving through empty space. It has made us treat vacuum as absence, matter as substance, gravity as external influence, and time as the arena in which matter slowly arranges itself. But perhaps this entire arrangement begins too late.

Perhaps before matter, there is field.

Before object, there is condition.

Before structure, there is alignment.

Before visibility, there is energy seeking stable expression.

This chapter begins from that deeper point.

Matter is not denied.

Matter is repositioned.

It is not the foundation of the universe. It is the universe becoming visible through field consequence. It is what appears when structured energy holds itself long enough, coherently enough, and intensely enough to be measured as form.

The universe is not made first of dead things.

It becomes visible through living field behavior.

And once matter is understood in this way, the so-called anomalies of structure begin to look very different.

4.1 — The Object-Based Habit of Human Thinking

Human thinking begins with objects.

This is not a weakness. It is how the mind first learns reality. Before a child understands energy, field, pressure, vibration, temperature, or force, the child understands things. A toy is a thing. A wall is a thing. A stone is a thing. A chair is a thing. A body is a thing. The world first arrives as shapes, surfaces, weights, colors, sounds, and names.

This habit becomes deeply rooted.

We point to something and name it. Once it has a name, we treat it as an independent identity. This is a cup. This is a tree. This is a star. This is a planet. This is a galaxy. The naming gives mental stability. The object becomes easier to remember, describe, compare, measure, and use.

Without this object-based thinking, ordinary life would be impossible.

A farmer cannot plough a field by thinking only of field dynamics. A carpenter cannot build a table by ignoring wood as material. A doctor cannot treat the body without recognizing organs, tissues, bones, and cells. An engineer cannot design a bridge without treating steel, concrete, load, support, and structure as practical realities.

So object-based thinking is useful.

It is necessary.

It is efficient.

But the problem begins when usefulness is mistaken for final truth.

The object we see may not be the deepest reality of what exists. It may be only the stable appearance of a process. It may be a pattern held together by conditions. It may be a temporary arrangement that appears permanent because its internal change is slower than our observation.

A stone feels solid because its structure resists our touch. But at deeper levels, it is not dead stillness. It is arrangement, binding, vibration, interaction, and energy stability. A cloud appears like an object in the sky, but it is not a solid thing. It is a temporary condensation of water vapor under atmospheric conditions. A flame appears as a glowing shape, but it is not an object sitting independently in space. It is a process maintained by fuel, oxygen, heat, and chemical transformation.

The mind sees "thing."

Reality may be showing "process."

This gap is small in daily life but enormous in cosmology.

When we look at the universe, we carry the same object-based habit. A star becomes a thing. A planet becomes a thing. A galaxy becomes a thing. Even a black hole becomes a thing. Then science begins to ask how these things move, attract, grow, collide, merge, and evolve.

This way of thinking is powerful because it allows measurement. Objects can be counted. Mass can be estimated. Distance can be calculated. Motion can be tracked. Structure can be classified. Images can be compared. A galaxy can be treated as a measurable object with size, brightness, mass, redshift, and composition.

But the danger is subtle.

Once we treat the galaxy as a thing, we may forget that the galaxy is also a field condition. Once we treat matter as substance, we may forget that substance may be stabilized energy. Once we treat mass as possession, we may forget that mass may be an expression of field density, resistance, and interaction.

The object becomes too convincing.

It hides the process that sustains it.

This object-based habit also shapes how we imagine formation. If a galaxy is a thing, then it must be built like a thing. Matter must gather. Parts must assemble. Stars must collect. Structure must grow over time. This leads naturally to the slow mechanical story of cosmic evolution. In that story, galaxies are large objects built from smaller objects through long processes.

Again, this is not useless.

But it may be incomplete.

Universal Energy Dynamics asks us to move behind the object. It asks us to see matter not merely as a thing but as a field consequence. It asks us to see form not as independent identity but as stabilized expression. It asks us to see the visible universe not as a warehouse of objects but as a structured field of energy manifestations.

This does not deny objects.

It explains why objects appear.

A wave is real, but it is not separate from the ocean. A whirlpool is real, but it is not an independent substance. A cloud is real, but it is not fixed identity. A flame is real, but it is not a solid thing. It exists as long as the sustaining conditions remain active.

Matter may be understood in a similar way.

A particle, a star, a planet, or a galaxy may appear as an object because the underlying energy configuration is stable enough to hold form. The object is the surface identity. The field condition is the deeper reality.

This is why UED uses the language of ripple condensation.

Matter is not simply a thing placed inside empty space. It is a local stabilization of ripple behavior within Structured Vacuum Energy. It appears as form when energy becomes condensed, coherent, and measurable. Its identity depends on the field conditions that sustain it.

This shift is difficult because our senses are object-trained.

We do not touch field consequence directly. We touch the resistance of form. We do not see ripple condensation directly. We see matter. We do not experience Structured Vacuum Energy through ordinary perception. We experience its consequences as objects, forces, motion, light, and structure.

So the mind naturally stops at the object.

But science must not stop where the senses stop.

The senses show us the readable surface. Intelligence must ask what makes that surface possible. The eye sees the star. The mind must ask what field process sustains the star. The telescope sees the galaxy. Interpretation must ask what deeper energy condition produces galactic structure.

This is the real challenge.

The universe may not be made of objects in the way daily life suggests. It may be appearing as objects because deeper energy fields have stabilized into measurable patterns. What we call matter may be the universe's way of becoming locally visible.

The object-based habit helped humanity survive, build, measure, and classify.

But now, at cosmic scale, it may be limiting our interpretation.

JWST shows us galaxies that appear too mature for the timeline. The object-based mind asks how those objects gathered enough matter so quickly. UED asks whether those objects are better understood as field consequences. The first question remains trapped in collection. The second question opens the possibility of condensation, alignment, and structural emergence.

This is why the habit must be examined.

An object is not false.

But an object is not final.

It is the visible face of a deeper process.

4.2 — The Limits of Matter-First Cosmology

When matter is treated as the starting point, the universe naturally becomes a construction story.

This construction story looks simple at first. Matter exists. Matter moves. Matter attracts. Matter gathers. Matter forms stars. Stars gather into galaxies. Galaxies form clusters. Clusters form large-scale structure. Everything begins with material presence and then develops through interaction over time.

This is the matter-first view of cosmology.

It is powerful because it gives the mind something concrete to hold. Matter feels real. Matter can be measured. Matter appears to carry mass, position, motion, temperature, luminosity, and structure. So it becomes tempting to say that the universe is made of matter, and everything else is merely the behavior of matter.

But this is where the limitation begins.

If matter is the first reality, then every cosmic structure must be explained by the movement and accumulation of matter. A galaxy must be explained by gas clouds gathering, stars forming, smaller structures merging, and gravitational attraction organizing scattered material over vast periods. The question always becomes: how did enough matter collect, and how long did it take?

This approach makes time compulsory.

Time becomes the silent builder behind cosmic structure. If a galaxy is large, it must have had enough time to gather. If it is bright, it must have had enough time to form enough stars. If it is chemically complex, it must have had enough time for earlier stars to live, die, and enrich the surroundings. If it is compact and organized, it must have had enough time for gravitational settling and structural development.

Within this logic, maturity belongs to duration.

This is why JWST creates discomfort.

When JWST observes galaxies and compact structures at high redshift that appear bright, dense, organized, or massive, the matter-first model immediately faces a time problem. If matter needs long time to collect and organize, then how did such structures appear so soon within the assigned cosmic timeline?

The model then searches for acceleration.

Perhaps star formation was more efficient. Perhaps gas collapsed faster. Perhaps dark matter halos formed earlier. Perhaps mergers happened rapidly. Perhaps black holes contributed to brightness. Perhaps stellar populations were unusual. Perhaps observational estimates require correction.

These are valid scientific possibilities. They should be tested carefully. But they still remain inside the same matter-first grammar. They do not question whether the structure was correctly understood as a slow collection of matter in the first place. They only ask how the collection happened faster.

That is the limitation.

A matter-first cosmology can become trapped inside the question of accumulation.

It keeps asking where the matter came from, how it gathered, how it cooled, how it collapsed, how it formed stars, how it merged, and how it produced visible structure. These are important questions, but they may not be the deepest questions. Beneath them lies a more fundamental one: what is matter itself?

If matter is not primary, then the whole formation story changes.

If matter is a field consequence, then cosmic structure is not merely the organization of already existing substance. It is the manifestation of structured energy into measurable form. A galaxy is not only a collection of particles. It is a large-scale stabilization of field behavior.

This is where Universal Energy Dynamics shifts the foundation.

In UED, matter is not ignored, but it is repositioned. It is no longer the first actor. It is the visible outcome of deeper energy dynamics. What appears as matter is ripple condensation within Structured Vacuum Energy. What appears as mass is a measurable expression of energy density, resistance, and field stability. What appears as structure is not merely collected substance, but aligned field consequence.

This changes the question from accumulation to manifestation.

Instead of asking only, "How did matter gather so quickly?" UED asks, "What field condition allowed matter to condense into this structure?"

Instead of asking only, "How did the galaxy become massive so early?" UED asks, "What energy density and ripple alignment produced this measurable mass?"

Instead of asking only, "How did the object mature within limited time?" UED asks, "Is maturity being misread as age, when it may actually indicate field coherence?"

The matter-first model struggles because it treats visible matter as the starting inventory of the universe. But if the visible inventory itself is a consequence of deeper field behavior, then cosmic structure may not require the same slow mechanical pathway imagined by material accumulation alone.

This also affects the meaning of gravity.

In matter-first cosmology, matter and mass appear first, and gravity then acts upon them. Gravity becomes the great organizer of material structure. But in UED, gravity is not the first cause of structure. Gravity itself is an emergent consequence of energy gradients seeking balance. If matter is already a field condensation, then gravity is not an external commander acting on dead substance. It is part of the same energy-field behavior through which structure becomes stable.

The difference is subtle but enormous.

In one view, matter exists first and gravity organizes it.

In the UED view, structured energy behavior produces matter, mass, gravity, and structure as connected consequences.

This removes the artificial separation between object and field.

Matter-first cosmology also depends heavily on the idea of empty space. Matter is imagined as something inside space. Objects move through space. Galaxies expand away from one another through space. Light travels across space. But if the vacuum itself is structured, then space is not a passive container. It is part of the reality through which matter becomes possible.

This is why matter-first cosmology may misread deep observations.

It sees a bright high-redshift galaxy and asks how enough material gathered there in such little time. But UED sees a field condition and asks how Structured Vacuum Energy condensed into such coherent expression. The first interpretation sees a construction delay. The second sees an alignment threshold.

A construction delay creates anomaly.

An alignment threshold creates consequence.

This is the deeper reason the "early galaxy" problem appears so serious in the standard model. It is not only because the galaxies are bright or massive. It is because matter-first thinking has already decided that such brightness and mass require long material history. The anomaly arises when observation refuses that required history.

But perhaps the requirement itself is incomplete.

The universe may not be assembling itself only from dead pieces across time. It may be continuously manifesting structure through field conditions. Matter may be the visible state of energy that has crossed into stability. Galaxies may be large-scale expressions of that stability. Little Red Dots may be compact high-tension condensation zones rather than immature objects that grew too quickly.

This does not remove the need for measurement.

It changes what measurement means.

Mass, luminosity, redshift, density, and structure remain important. But they should not be interpreted only through the lens of material accumulation. They may also be read as signatures of field condition, ripple saturation, energy density, and alignment behavior.

That is why matter-first cosmology has limits.

It can describe the visible arrangement of matter, but it may not fully explain the origin of matter as structured consequence. It can model how objects interact, but it may miss the deeper field condition that makes the object possible. It can calculate growth, but it may misunderstand manifestation.

The universe may not be made of matter first.

It may become matter when structured energy condenses into observable form.

And once this is understood, the question of cosmic structure changes permanently.

The galaxy is no longer only something matter built.

It is something the field expressed.

4.3 — Structured Vacuum Energy: The Forgotten Medium

The greatest misunderstanding of modern physics may not be about matter.

It may be about the vacuum.

We usually imagine the vacuum as emptiness. A blank background. A silent stage. A container in which objects move, light travels, stars shine, galaxies rotate, and forces act. In ordinary thinking, vacuum means absence. It means nothing is there. It means space without matter.

This imagination is understandable.

When we remove air from a chamber, we call it vacuum. When we look between stars, we imagine emptiness. When we see distance between galaxies, we imagine empty space separating them. The human mind naturally treats visible things as real and invisible background as empty.

But what if the vacuum is not nothing?

What if the so-called emptiness is the deepest condition of physical reality?

What if matter is not placed inside the vacuum, but emerges from the structured behavior of the vacuum itself?

This is where Universal Energy Dynamics begins to separate itself from the ordinary matter-first imagination.

In UED, the vacuum is not dead emptiness. It is Structured Vacuum Energy, or SVE. It is the silent field-medium through which energy becomes active, through which ripple behavior arises, through which matter condenses, through which gravity emerges, and through which observable reality becomes possible.

This does not mean the vacuum is a material substance in the ordinary sense. It is not dust. It is not gas. It is not invisible matter floating everywhere. It is a structured field condition, a deeper energy-state that allows manifestation to occur.

The word "structured" is important.

If the vacuum were truly nothing, it could not support anything. It could not carry interaction. It could not allow propagation. It could not respond to energy gradients. It could not permit stable matter formation. It could not produce measurable consequences. Absolute nothingness cannot act, resist, transmit, balance, or manifest.

But the universe acts.

Light propagates. Fields interact. Particles appear and disappear in quantum discussions. Gravity behaves across vast distances. Space itself is not observationally meaningless. The so-called empty regions of the universe are not dead absence. They participate in reality.

UED takes this participation seriously.

Structured Vacuum Energy is the forgotten medium because modern interpretation often talks about objects while quietly depending on the field that makes objects possible. We discuss galaxies, stars, black holes, matter, mass, and expansion, but we do not always ask what deeper condition allows these to exist as stable observable consequences.

The stage is treated as passive.

But perhaps the stage is not passive at all.

Perhaps the stage is the source of the drama.

A simple analogy may help. A wave on water cannot be understood by looking only at the wave as if it were an independent object. The wave is real, but it is not separate from water. Its height, motion, shape, and behavior depend on the medium. A whirlpool is also real, but it is not a thing added into water from outside. It is a structured motion within water. A cloud is visible, but it is not an independent solid object placed in the sky. It is condensation within atmospheric conditions.

Similarly, matter may be understood as condensation within the deeper field.

The vacuum is not a blank container.

It is the condition through which matter becomes possible.

This changes the meaning of cosmic structure. If matter is understood as something placed inside empty space, then galaxies must be explained as collections of that matter. But if matter is ripple condensation within Structured Vacuum Energy, then galaxies become field expressions. Their structure is not merely the result of collected substance; it is the visible outcome of field organization.

This also changes the meaning of energy.

Energy is not merely something possessed by objects. It is the operational basis of manifestation. It is the active capacity through which the field expresses form, motion, interaction, resistance, and balance. When energy remains stable and unexpressed, it may not appear as matter. When energy enters structured action within SVE, ripple behavior arises. When ripple behavior stabilizes into coherent condensation, matter becomes measurable.

In this sense, matter is not created from emptiness.

Matter is expressed from structured potential.

This is why SVE is essential to understanding JWST observations. If deep-field galaxies and Little Red Dots are interpreted only as matter objects sitting far away in a chronological past, they become difficult. Their brightness, compactness, and apparent maturity create problems for the slow mechanical timeline. But if they are understood as field consequences arising within SVE, the question changes.

We no longer ask only how matter gathered so quickly.

We ask what field tension produced such condensation.

We no longer ask only why such structures appear so early.

We ask what deeper field layer JWST is resolving.

We no longer ask only how compact objects became so energetic.

We ask whether compactness itself is a sign of localized field saturation and ripple alignment.

This is the power of reintroducing the medium.

A universe without a meaningful medium must depend heavily on particles, forces, spacetime geometry, and chronology. A universe with Structured Vacuum Energy allows matter, gravity, time, redshift, and structure to be understood as connected consequences of deeper field behavior.

This does not make the universe vague.

It makes it more integrated.

Modern science often fragments reality into separate categories: matter, space, time, gravity, energy, light, and particles. UED does not deny these categories, but it asks whether they are surface divisions of a deeper unity. SVE becomes the underlying condition from which these divisions become observable.

Matter is field condensation.

Gravity is field-gradient behavior.

Time is alignment delay.

Redshift is field-condition signature.

Structure is energy achieving coherence.

Anomaly is misread consequence.

This is not merely a change of vocabulary. It is a change of foundation.

If the vacuum is empty, matter must be explained as the main actor. But if the vacuum is structured, matter becomes one expression of a deeper field. If the vacuum is empty, cosmic distance becomes a passive gap. But if the vacuum is structured, cosmic distance may involve field resistance, ripple strain, and alignment behavior. If the vacuum is empty, redshift is mainly a stretching of light across expansion. But if the vacuum is structured, redshift may also carry the signature of the field through which the signal resolves.

This is why SVE is the forgotten medium.

It was always there, silently supporting every observation, but interpretation often looked past it. We measured objects, calculated distances, estimated masses, and built timelines, but the deeper medium was treated as background. UED brings it forward.

The universe is not a collection of objects floating in nothing.

It is a field reality expressing itself as objects.

Once this is understood, matter loses its false independence. It is still real, but it is no longer isolated. It is no longer the first brick of existence. It is the visible condensation of structured energy within the vacuum field.

This is where the UED reinterpretation becomes powerful.

The vacuum is not the absence of reality.

It is the silent condition from which reality becomes measurable.

4.4 — Matter as Ripple Condensation

Matter appears solid because the field has become stable enough to resist disturbance.

This is the first thing to understand.

When we touch a table, a stone, a wall, or our own body, we experience solidity. The senses tell us that matter is firm, fixed, and independent. It appears to exist by itself. It appears to have its own identity, its own boundaries, its own mass, and its own presence.

But what appears solid at one level may be dynamic at another.

A wave appears as a form, but it is not separate from the ocean. A whirlpool appears as a local structure, but it is not an object added into water. A cloud appears as a visible body in the sky, but it is a condensation of atmospheric conditions. A flame appears as a shape, but it is sustained by an active process. In all these cases, form is real, but it is not independent. It exists because a deeper medium carries a certain condition.

Universal Energy Dynamics applies this insight to matter itself.

Matter is not an independent substance sitting inside empty space.

Matter is ripple condensation within Structured Vacuum Energy.

This statement must be understood carefully. Ripple condensation does not mean matter is imaginary. It means matter is the stable expression of deeper field behavior. Energy within SVE does not always appear as matter.

It may remain unexpressed, distributed, balanced, or below the threshold of manifestation. But when field disturbance, energy density, resistance, and alignment conditions reach a suitable state, ripple behavior condenses into stable form. That stable form becomes observable as matter.

In this sense, matter is not the first layer of reality.

It is the visible layer of field resolution.

The field carries possibility. Energy activates the possibility. Ripple behavior expresses disturbance. Condensation stabilizes that disturbance into form. Matter is the form that becomes measurable when the field holds energy in coherent structure.

This is very different from the usual imagination.

In matter-first thinking, the universe is made of things. These things move, collide, attract, and build larger things. Matter is the raw material. The vacuum is the background. Gravity is the organizer. Time is the builder. Structure is the result.

In UED, the order changes.

The field comes first.

Structured Vacuum Energy provides the condition. Energy becomes active within that condition. Ripple behavior arises. The field seeks balance. Disturbance contracts toward equilibrium. Where the conditions allow coherence, matter condenses. Matter is therefore not raw material in the deepest sense. It is manifested stability.

A simple analogy is water vapor becoming cloud.

The cloud is visible. We can point to it. We can photograph it. We can describe its shape. But the cloud is not a permanent object independent of atmospheric condition. It appears when temperature, pressure, humidity, and condensation conditions permit water vapor to become visible as droplets. If those conditions change, the cloud changes or disappears.

Matter, under UED, is not exactly a cloud, but the analogy helps. Matter appears when the deeper field reaches the condition for stable manifestation. The visible object is not false. But its identity depends on the field process sustaining it.

Another analogy is the whirlpool.

A whirlpool has shape, motion, boundary, and identity. We can say, "There is a whirlpool." But no one thinks the whirlpool is a separate substance independent of water. It is a structured behavior of water under specific flow conditions. Remove the conditions, and the whirlpool vanishes. The water remains, but the form dissolves.

Matter can be viewed similarly.

The form appears. The deeper energy remains.

This is why the phrase "ripple condensation" is so powerful. It avoids two errors at once. It avoids the error of treating matter as dead, independent substance. It also avoids the error of dismissing matter as unreal. Matter is real as condensation. It is real as field structure. It is real as stable consequence. But it is not ultimate by itself.

It is the field becoming local.

It is energy becoming measurable.

It is structure becoming visible.

This understanding changes the meaning of mass.

If matter is ripple condensation, mass is not simply a fixed possession hidden inside an object. Mass becomes a measurable expression of how energy is condensed, stabilized, and resisted within the field. The more coherent and dense the condensation, the stronger the measurable presence. What we call mass may therefore be understood as field density under stable manifestation, not as an isolated property belonging to matter alone.

This also changes the meaning of identity.

A particle, a star, a planet, or a galaxy appears to have identity because its field configuration remains coherent for a certain duration. The identity is not false, but it is conditional. It remains as long as the structure remains. When the field configuration changes, identity transforms. A star collapses, a planet erodes, a particle interacts, a galaxy reorganizes. Matter is not frozen being. It is sustained pattern.

This is where UED becomes directly relevant to cosmic anomalies.

If matter is assumed to be collected dust, then large structures require long accumulation. But if matter is ripple condensation, then structure may emerge when field conditions reach threshold. A galaxy is not only a mass of collected stars. It is a large-scale condensation pattern within SVE. A Little Red Dot is not necessarily a tiny object that somehow gathered too much material too fast. It may be a compact ripple-condensation zone where field tension, energy density, and resistance have produced intense structural expression.

This makes the JWST observations less shocking.

The shock exists because we expected matter to assemble slowly. But if matter itself is the result of field condensation, the deeper question is not only how much matter assembled. The deeper question is what field condition condensed into the appearance of matter at that scale and intensity.

The mainstream model sees a mature deep galaxy and asks, "How did so much matter gather so early?"

UED asks, "What field state produced such coherent matter-expression?"

The mainstream model sees compact brightness and asks, "What mechanism made it shine so much in so little time?"

UED asks, "What ripple tension and energy-density condition generated such luminous condensation?"

The mainstream model sees mass as inventory.

UED sees mass as field consequence.

This does not eliminate measurement. It gives measurement a deeper context. Mass, brightness, density, and structure remain important. But they are no longer treated as properties of isolated things. They become signs of how the field has condensed and stabilized energy into observable form.

Once this is understood, the universe appears less like a warehouse and more like a living field of expression.

Matter is not floating inside vacuum.

Matter is the vacuum field expressing itself under structured energy conditions.

This is the foundation of the UED reinterpretation. Without this shift, galaxies remain collections, black holes remain objects, gravity remains attraction, and time remains the builder. With this shift, galaxies become field consequences, black holes become saturation zones, gravity becomes gradient behavior, and time becomes alignment delay.

Matter is where the illusion begins.

And matter is where the reinterpretation must begin.

Because if matter is not a thing, but a field consequence, then the entire universe is not made of dead objects.

It is made visible through structured energy becoming form.

4.5 — Mass, Identity, and Stability as Field Conditions

Mass feels like possession.

We speak as if an object has mass in the same way it has a name, a shape, or a location. A stone has mass. A planet has mass. A star has mass. A galaxy has mass. The language is so familiar that we rarely pause to question it.

But language can quietly mislead us.

When we say an object "has" mass, the mind begins to imagine mass as something stored inside the object. It appears like a fixed property, a permanent possession, an internal weight of existence. The object becomes the owner, and mass becomes the owned quality.

Universal Energy Dynamics asks us to look deeper.

If matter itself is ripple condensation within Structured Vacuum Energy, then mass cannot be understood merely as an independent possession of matter. Mass must be understood as a measurable expression of field condition. It is how condensed, stable, resistant, and coherent a particular energy configuration has become within the field.

In simple words, mass is not just what matter has.

Mass is how the field holds matter.

This is a major shift.

In ordinary thinking, mass belongs to the object. In UED, mass belongs to the relationship between energy condensation, field stability, resistance, and interaction. The object appears to have mass because the field configuration has become stable enough to resist change and measurable enough to behave as matter.

A stone feels heavy not because it is an isolated thing carrying some mysterious inner burden, but because its field-structured condensation interacts with the surrounding field and with our body in a measurable way. A planet has gravitational influence not because it is merely a lump of material sitting in empty space, but because its condensed energy configuration creates a field-gradient relationship with the surrounding Structured Vacuum Energy.

The more stable the condensation, the stronger the measurable identity.

The more intense the energy density, the stronger the field presence.

The more coherent the structure, the more persistent the object appears.

This is why mass, identity, and stability cannot be separated.

An object appears as an object because its field configuration remains coherent. If that coherence breaks, the object changes. If the structure dissolves, the identity disappears. If the field condition transforms, the matter-form transforms with it.

Nothing in the universe is an isolated permanent thing.

Everything is sustained condition.

This does not mean objects are fake. A mountain is real. A planet is real. A star is real. A human body is real. A galaxy is real. But their reality is conditional. They are real as stabilized patterns, not as independent absolutes. They exist as long as the energy configuration supporting them remains coherent within the field.

A whirlpool has identity while the water maintains that motion. A flame has identity while the conditions sustain combustion. A cloud has identity while atmospheric conditions hold condensation. Similarly, matter has identity while the field holds energy in stable form.

This understanding helps us avoid one of the biggest errors in cosmology: treating cosmic objects as if they are self-contained things.

A star is not merely a hot ball of gas. It is a field-supported energy structure. A galaxy is not merely a collection of stars. It is a large-scale field arrangement. A black hole candidate is not necessarily a singular object of infinite density. It may be a saturation condition where field coherence collapses beyond ordinary observation.

Once mass is understood as field condition, the idea of fixed intrinsic identity begins to weaken.

What is a galaxy?

In the matter-first view, it is a collection of stars, gas, dust, dark matter, and gravitational structure. In UED, these descriptions may still be useful at the surface level, but they are not final. A galaxy is a coherent large-scale field configuration where energy has condensed, stabilized, rotated, radiated, and organized into measurable structure.

Its identity is not merely in its visible stars.

Its identity is in its field coherence.

This is why a galaxy can maintain shape, motion, and gravitational behavior across enormous scales. It is not simply a crowd of objects accidentally held together. It is a structural field consequence. Its apparent unity comes from the coherence of energy gradients, ripple alignments, and equilibrium-seeking behavior across the region.

The same applies to Little Red Dots.

If we treat them as small objects carrying fixed mass and brightness, we immediately ask how such mass and luminosity could exist so early. But if we treat them as compact field conditions, the question changes. Their mass estimate becomes a sign of condensed energy density. Their brightness becomes a sign of energetic release or field tension. Their compactness becomes a sign of strong local coherence. Their redness may carry information about field resistance and observational alignment, not merely distance.

This does not remove the need for measurement.

It deepens measurement.

Mass remains measurable. Brightness remains measurable. Redshift remains measurable. But these measurements are not final labels. They are clues about the underlying field condition.

In this sense, mass is like the visible pressure of a hidden process.

When a region of SVE condenses energy into stable form, that form becomes measurable as matter. When the condensation is strong, it expresses stronger mass. When the field coherence persists, the object appears stable. When the configuration interacts with other configurations, gravity-like behavior emerges as energy gradients seek balance.

This also explains why cosmic structures can appear more mature than expected. Maturity, in UED, is not merely age. It is coherence. A field structure can be highly coherent if the conditions allow it, even if the chronological interpretation assigns it an early position. A galaxy may appear mature not because it has lived through a long mechanical childhood, but because its field arrangement is already stable and intense.

This is a key correction.

The old model treats maturity as accumulated history.

UED treats maturity as achieved stability.

The difference is enormous.

If maturity is accumulated history, then high-redshift mature galaxies become anomalies. If maturity is achieved field stability, then the same galaxies become signals of intense coherent field formation.

The universe may not be asking, "How old is this object?"

It may be asking, "How stable is this configuration?"

This also changes the meaning of cosmic identity. We often speak as if the universe contains separate things: galaxies here, stars there, black holes somewhere else, empty space between them. But under UED, separation is a surface appearance. At the deeper level, all forms arise within the same structured field. Each object is a local condition of that field. Each identity is a temporary coherence within universal energy behavior.

The object is local.

The field is continuous.

The identity is conditional.

This does not reduce the grandeur of the universe. It increases it. A galaxy becomes more than a collection. A star becomes more than a furnace. Matter becomes more than substance. Each becomes a readable expression of deeper energy intelligence.

This is why the concept of fixed intrinsic identity must be handled carefully. Matter appears to have fixed identity because the field condition remains stable over the timescale of observation. But when the timescale changes, identity changes. A star that appears stable for millions of years may eventually collapse. A planet that appears permanent may gradually transform. A galaxy that appears coherent may merge, reorganize, or dissolve. Stability is relative to the duration and scale of observation.

The universe is not made of permanently fixed objects.

It is made of sustained field arrangements.

This is also why anomalies arise when we treat objects too rigidly. If we assume a galaxy has fixed identity as a matter-collection, we judge it by how long that matter-collection should take to form. But if we understand it as a field condition, we judge it by coherence, energy density, and alignment threshold.

One view creates a time problem.

The other reveals a field process.

In UED, mass is not the king of explanation. It is one expression among others. Mass, brightness, motion, compactness, redshift, and structure are all signs of how the field is behaving. They are not separate mysteries. They are connected outputs of deeper energy dynamics.

Matter becomes visible.

Mass becomes measurable.

Identity becomes stable.

Gravity becomes emergent.

Structure becomes coherent.

All because the field holds energy in a particular condition.

This is why matter cannot be treated as an independent substance with fixed identity. It is not a dead thing carrying permanent selfhood. It is a consequence of field stabilization. Its mass is not merely possessed; it is expressed. Its identity is not absolute; it is sustained. Its stability is not guaranteed; it is conditional.

And once this is understood, the universe becomes easier to reinterpret.

Galaxies are no longer impossible structures arriving too early.

They are coherent field configurations being read through a limited timeline.

Little Red Dots are no longer embarrassing compact objects.

They are possible high-density field expressions.

Black holes are no longer monsters of infinite matter.

They are likely saturation zones where field coherence reaches a boundary.

The entire cosmic picture shifts.

Matter is not the final actor on the stage.

Matter is the stage becoming visible in local form.

4.6 — Why This Changes the Meaning of Cosmic Structure

Once matter is understood as a field consequence, cosmic structure can no longer be interpreted in the old way.

This is the turning point.

If matter is treated as an independent substance, then every structure in the universe becomes a material arrangement. A star becomes collected gas. A planet becomes condensed dust and rock. A galaxy becomes a large gathering of stars, gas, dust, and dark matter. A black hole becomes a collapsed object of extreme mass. The universe becomes a giant warehouse of things moving, attracting, colliding, merging, and evolving through time.

This view appears practical because it works at the surface level.

We can measure stars. We can estimate mass. We can observe galaxies. We can track motion. We can compare brightness. We can classify structures. So the object-based interpretation seems natural. But when observations become sharper, the limits of this interpretation begin to appear.

JWST has exposed these limits.

It has shown compact, bright, mature-looking structures at cosmic depth. If these are interpreted only as matter-collections, then the question becomes difficult: how did enough matter gather, organize, shine, and stabilize so early? Why does structure appear where the timeline expected incompleteness? Why does compact maturity appear where slow mechanical evolution expected primitive formation?

The old model must answer by speeding up the construction story.

Faster star formation.

Faster collapse.

Earlier black hole seeds.

More efficient gas conversion.

Hidden scaffolding.

Special conditions.

All these may be technically explored, but they remain inside the same assumption: cosmic structure is built by matter gathering over time.

Universal Energy Dynamics changes the question.

If matter itself is ripple condensation within Structured Vacuum Energy, then cosmic structures are not merely built from matter. They are expressed through field conditions. A galaxy is not only a crowd of stars. It is a large-scale field consequence. A Little Red Dot is not only a compact object with puzzling brightness. It may be a localized zone of intense ripple condensation. A black hole candidate is not only an ultra-dense object. It may be a saturation threshold where field coherence collapses from the standpoint of observation.

This does not deny the visible structure.

It explains it differently.

The galaxy still has stars. The stars still radiate. The gas still moves. The mass is still measured. The redshift is still recorded. But these are not treated as isolated facts belonging to separate objects. They are read as signs of field behavior.

Mass becomes a clue.

Brightness becomes a clue.

Compactness becomes a clue.

Redshift becomes a clue.

Motion becomes a clue.

Structure becomes a clue.

Each tells us something about the condition of the field that produced and sustains the observable form.

This is a deeper way of reading the cosmos.

In the matter-first view, cosmic structure is mostly a historical result. It asks: what happened over time to build this object? In the UED view, cosmic structure is also a present field condition. It asks: what energy configuration is being expressed here? What kind of field tension is active? What level of ripple alignment has been achieved? What resistance is shaping the signal? What equilibrium process is visible through this form?

This changes the interpretation of maturity.

A mature galaxy, in the old view, means a galaxy that has undergone enough historical development. It must have had enough time to collect matter, form stars, enrich its environment, stabilize its motion, and organize its structure. So if such maturity appears at high redshift, it becomes a chronological problem.

But in UED, maturity is not only age.

Maturity is coherence.

A structure is mature when its field configuration is coherent enough to appear stable, bright, compact, organized, or self-sustaining. That coherence may arise through field alignment, not only through long mechanical history. This does not mean time is irrelevant. It means time is not the only builder. Alignment, density, resistance, and equilibrium-seeking behavior also matter.

This is why the meaning of cosmic structure changes completely.

The old question is: how did this structure form so early?

The UED question is: what field condition made this structure visible?

The old question is: how did matter assemble so fast?

The UED question is: how did energy condense into this form?

The old question is: why does the galaxy look too mature?

The UED question is: why are we measuring maturity only by age?

This shift helps us understand why cosmic anomalies arise. Many anomalies are not caused by the observation itself. They are caused by forcing the observation into a matter-first, time-dependent framework. Once the framework changes, the same observation may no longer look anomalous.

A Little Red Dot, for example, appears troubling because it is compact, red, bright, and located at cosmic depth. In the standard reading, it must be explained within the early-universe timeline. But under UED, it may be read as a compact field-expression, a high-density condensation zone where Structured Vacuum Energy has reached a local condition of intense manifestation. Its brightness may not only require unusual stellar history. It may reflect concentrated field realignment. Its redness may not only indicate distance. It may also carry information about resistance, strain, and signal resolution through the field.

Similarly, a massive-looking high-redshift galaxy is not automatically a structure that collected too much matter too quickly. It may be a large-scale coherence pattern where matter appears as the local expression of deeper energy condensation. Its mass estimate becomes not merely an inventory count, but a field-density signature.

This is why matter as field consequence is not a minor philosophical adjustment.

It changes the entire diagnostic method of cosmology.

When a doctor examines a patient, symptoms are not treated as separate events. Fever, pain, fatigue, inflammation, and pulse are read together as signs of underlying condition. Similarly, UED asks cosmology to stop treating mass, brightness, redshift, compactness, and structure as separate puzzles. They may be connected symptoms of the field.

The cosmos may not be a collection of independent objects.

It may be a continuous field expressing local consequences.

Once this is accepted, galaxies are no longer isolated islands floating in empty space. They are field configurations. Their shapes, motions, luminosities, and central regulators belong to the same deeper energy behavior. Clusters are not merely groups of galaxies. They are larger-scale field arrangements. Cosmic webs are not merely matter distributed across space. They may be the visible geometry of structured energy seeking balance.

This also prepares the reinterpretation of gravity.

If matter is not an independent thing, then gravity cannot be understood only as attraction between independent things. Gravity must be connected to the field condition that allows matter and mass to appear. It becomes less like a force pulling objects from outside and more like an emergent consequence of energy gradients, resistance differences, and equilibrium-seeking behavior inside the field.

This is why Chapter 4 must come before the chapter on gravity.

Matter must be repositioned first.

Only then can gravity be reinterpreted properly.

If matter is the beginning, gravity becomes the organizer of matter.

If matter is consequence, gravity becomes part of the same field process through which matter appears, stabilizes, and interacts.

That is a very different universe.

It is not a universe of dead things waiting to be moved.

It is a universe of field conditions becoming visible as things.

This is the deeper meaning of cosmic structure under UED. Structure is not only arrangement. Structure is expression. It is the field showing its state. It is energy becoming readable. It is equilibrium attempting to resolve disturbance through form.

A galaxy is not merely built.

It is expressed.

A star is not merely burning matter.

It is a field condition radiating through condensed energy.

A planet is not merely material gathered into a sphere.

It is stable field condensation operating within larger gradients.

A black hole candidate is not merely mass collapsed into darkness.

It may be a saturation boundary where observable coherence fails.

Once matter is understood this way, the universe becomes more unified. The same principles begin to apply across scale. Small structures and large structures are not separate mysteries. They are different expressions of field behavior under different conditions.

This is why UED has the power to reinterpret cosmic anomalies.

It does not patch one anomaly at a time. It changes the foundation from which anomalies are produced. If matter is not a thing but a field consequence, then the maturity of galaxies, the compactness of Little Red Dots, the behavior of black hole candidates, and the meaning of redshift all need to be read again.

The universe may not be asking us to explain how dead matter built impossible structures.

It may be asking us to understand how structured energy becomes visible as form.

And that is a much deeper question.

Closing Reflection — The Universe Is Not Made of Dead Things

The universe is not made of dead things.

This statement may sound simple, but it quietly changes everything.

For too long, human imagination has treated the cosmos as a grand warehouse of objects. Stars are placed here. Galaxies are placed there.

Planets move around stars. Black holes sit at centers. Matter occupies space. Light travels across distance. Gravity acts between masses. Time carries everything forward.

This picture is familiar.

It is useful.

But it may not be deep enough.

Because once matter is treated as dead substance, the universe becomes a mechanical arrangement of things. Then every cosmic structure must be explained through movement, collision, attraction, accumulation, and duration. Galaxies become crowds of stars. Stars become burning gas. Planets become gathered material. Black holes become collapsed matter. Space becomes the empty container in which all these objects sit.

But Universal Energy Dynamics asks us to look beneath this surface.

Matter is not the first truth.

Matter is what appears when structured energy becomes stable enough to be observed as form.

A stone is not merely a thing. A star is not merely a thing. A galaxy is not merely a thing. Each is a visible consequence of deeper field behavior. Each is energy held in a condition. Each is a local coherence within Structured Vacuum Energy. Each exists because the field has reached a state where ripple behavior can condense, stabilize, and become measurable.

This does not make matter unreal.

It makes matter more meaningful.

Matter is not an illusion to be dismissed. It is a consequence to be understood. It is the universe becoming visible. It is the field taking form. It is energy entering a condition of stability, resistance, and identity. It is not dead material lying inside emptiness. It is structured expression.

This shift is necessary because many cosmic anomalies arise from treating matter too rigidly.

If matter is fixed substance, then galaxies must be built slowly. If mass is independent possession, then massive galaxies at cosmic depth become difficult. If structure is collected material, then compact maturity becomes a problem. If vacuum is empty, then the medium through which matter appears is forgotten. If gravity is only the organizer of matter, then the deeper field process remains hidden.

But if matter is ripple condensation within Structured Vacuum Energy, the picture changes.

A galaxy is no longer merely a pile of stars.

It is a large-scale field consequence.

A Little Red Dot is no longer merely a puzzling compact object.

It may be a high-density zone of ripple condensation.

Mass is no longer only inventory.

It becomes a field-density signature.

Brightness is no longer only the output of accumulated stellar history.

It may also reveal energetic release from concentrated field alignment.

Maturity is no longer only age.

It may be coherence.

This is why Chapter 4 is the foundation of the UED reinterpretation.

Before we reinterpret gravity, black holes, redshift, or cosmic time, we must first reinterpret matter. If matter remains a dead thing, then all explanations remain trapped inside the old mechanical story. But once matter becomes field consequence, the universe opens differently. What looked impossible under accumulation may become natural under condensation. What looked too early under chronology may become meaningful under alignment. What looked anomalous under matter-first thinking may become expected under field-based interpretation.

The universe is not a construction site of dead parts.

It is a field of active consequences.

This does not mean every mystery is solved. It means the starting point changes. And when the starting point changes, old problems do not always remain in the same form. Some problems dissolve. Some become clearer. Some demand new language. Some reveal that the difficulty was not in reality, but in the way reality was being framed.

Matter was never the silent brick of the universe.

Matter is the visible grammar of the field.

It is how the unseen becomes seen.

It is how energy becomes readable.

It is how structure appears from condition.

It is how the universe presents itself to observation.

This is why the vacuum cannot be treated as absence. If matter is ripple condensation, then the medium matters. The field matters. The structured vacuum matters. Nothing visible can be understood properly if the invisible condition behind visibility is ignored.

The old imagination begins with matter and asks how it moves.

UED begins with the field and asks how matter appears.

That is the decisive shift.

Once this is understood, the universe no longer looks like a collection of objects floating in empty space. It looks like a continuous field expressing itself through local structures. Objects are real, but they are not final. Matter is real, but it is not primary. Mass is real, but it is not isolated. Identity is real, but it is conditional. Stability is real, but it is sustained by field coherence.

This chapter therefore leaves us with a new responsibility.

We must stop asking only what matter does.

We must ask what matter is.

And when we ask deeply enough, matter no longer appears as the beginning of reality. It appears as the visible consequence of a deeper reality.

The universe is not made of dead things.

It becomes visible through living field behavior.

Matter is not what the universe is made of.

Matter is how the universe shows itself.

Chapter 5 — Gravity Without the Myth of Spacetime Fabric

Gravity as an emergent consequence of energy gradients seeking balance, rather than the bending of empty spacetime.

The Most Trusted Illusion

Gravity is the oldest intimacy between the human body and the universe.

Before we learn the word, we experience it. A child falls before understanding falling. A stone drops before anyone explains motion. Rain comes down. Rivers flow downward. Bodies stay grounded. Planets remain in orbit. Stars hold worlds around them. Galaxies rotate across unimaginable scales.

Gravity is everywhere.

And because it is everywhere, we think we understand it.

But familiarity is not understanding.

This is one of the most powerful traps in science. The things we experience daily often become the things we question least. We do not wake up every morning astonished that our feet touch the ground. We do not pause before every falling object and ask what the universe is truly doing. We accept gravity as a fact of life because life would be impossible without accepting it practically.

But practical acceptance is not final explanation.

Newton gave gravity mathematical discipline. He described it with extraordinary clarity as an attraction between masses. His framework allowed humanity to calculate motion, predict orbits, understand planetary behavior, and build a mechanical picture of the cosmos. For its time, it was not merely successful; it was revolutionary.

Then Einstein changed the language.

Gravity was no longer simply a force pulling objects across empty space. It became the curvature of spacetime. Matter told spacetime how to curve, and curved spacetime told matter how to move. This was a magnificent shift. It gave gravity geometric elegance. It explained observations Newtonian gravity could not fully capture. It transformed the universe from a mechanical stage into a relational structure.

But even Einstein's explanation came with a danger.

Not necessarily in the mathematics, but in the imagination built around it.

To explain curved spacetime, we often use the metaphor of a fabric. A heavy ball placed on a rubber sheet creates a depression. Smaller balls roll toward it, not because they are pulled by an invisible rope, but because the surface itself is curved. This image is simple, beautiful, and useful for teaching.

But it is also dangerous.

Because slowly, the metaphor begins to harden into belief.

The mind begins to imagine spacetime as a real cloth-like thing stretched across the cosmos. We begin speaking as if empty spacetime is a fabric that bends, stretches, ripples, and carries objects. The metaphor, which was meant to help understanding, becomes a substitute for deeper questioning.

But what exactly is bending?

If spacetime is empty, how does emptiness curve?

If it is a fabric, what is it made of?

If it can ripple, respond, carry disturbance, and guide motion, can it truly be nothing?

This is where Universal Energy Dynamics enters the discussion.

UED does not deny the observed effects attributed to gravity. Bodies fall. Planets orbit. Light bends. Galaxies rotate. Massive structures influence motion. These observations are real. The question is not whether gravitational behavior exists. The question is whether the explanation has stopped too early.

Perhaps gravity is not a force pulling matter.

Perhaps it is not the bending of an empty fabric.

Perhaps gravity is the visible consequence of energy gradients seeking balance within Structured Vacuum Energy.

This shift matters because Chapter 4 has already changed the foundation. Matter is not an independent dead thing. Matter is ripple condensation within the structured field. Mass is not merely a fixed possession of an object. Mass is a measurable expression of energy density, coherence, resistance, and field stability.

If matter itself is field consequence, then gravity cannot be understood as something that begins only after matter appears.

Gravity must be part of the same field process.

Where energy condenses, the field condition changes. Where field condition changes, gradients appear. Where gradients appear, imbalance seeks resolution. What we call gravity may be the observable behavior of this resolution.

In this view, gravity is not an isolated command acting from one object upon another.

It is relational field behavior.

A planet does not pull a stone like a hand pulling a rope. A star does not hold planets merely by invisible attraction. A galaxy does not maintain structure only because matter commands space to bend. These are surface descriptions. At the deeper level, energy gradients inside Structured Vacuum Energy create directional tendencies toward balance. Motion follows the logic of field imbalance.

Gravity, then, becomes less mysterious and more integrated.

It is not a separate power injected into the universe.

It is the field revealing where equilibrium is unfinished.

This does not make Newton useless. It does not make Einstein meaningless. Both gave humanity powerful languages within their frames. Newton gave calculation. Einstein gave geometry. But UED asks for a deeper mechanism beneath both attraction and curvature.

If gravity is described as attraction, we still must ask what creates the attraction.

If gravity is described as curvature, we still must ask what physically underlies the curved behavior.

If spacetime behaves like a medium, perhaps the medium is not spacetime fabric at all.

Perhaps it is Structured Vacuum Energy.

This chapter begins from that possibility.

Gravity is the most trusted illusion because we trust its effects before we understand its origin. We trust falling, orbiting, bending, rotating, clustering, and collapsing. Then we attach familiar language to these behaviors and feel satisfied.

But the universe may not be pulling things together.

It may be balancing itself.

Gravity may not be the master force of matter.

It may be the signature of imbalance within the field.

And once gravity is understood in this way, the entire cosmic structure must be read again. Galaxies, black holes, redshift, dark matter, cosmic expansion, and the so-called anomalies seen by JWST all begin to look different.

Because the question is no longer only: how much mass is there?

The deeper question becomes: what gradient is the field trying to resolve?

5.1 — Why Gravity Became a Fabric in the Human Imagination

Every great theory eventually needs a picture.

Mathematics can convince the trained mind, but imagination needs an image. Equations may carry precision, but most people do not live inside equations. They understand through comparison, analogy, shape, movement, and familiar experience. So when a theory becomes difficult, the human mind searches for something it can see.

This is how gravity became a fabric.

Einstein's general theory of relativity changed the old idea of gravity as a simple invisible pull between masses. Instead of imagining the Sun pulling Earth by an unseen rope, the new picture suggested that mass and energy affect the geometry of spacetime, and objects move along the paths created by that geometry. This was a powerful intellectual shift. Gravity was no longer merely a force acting across distance. It became a relationship between matter, motion, and the structure of spacetime itself.

But such an idea is not easy to visualize.

How does one imagine spacetime? Space itself is difficult enough. Time is even more abstract. Combining space and time into a four-dimensional structure is beyond ordinary sensory experience. So the public imagination needed help. Teachers, writers, documentaries, and popular science explanations began using the famous rubber-sheet analogy.

Place a heavy ball on a stretched fabric.

The fabric bends.

Now roll a smaller ball nearby.

The smaller ball curves toward the heavier one, not because the bigger ball directly pulls it, but because the surface beneath it has changed shape.

This analogy became enormously successful.

It gave the mind something visible. It turned abstract geometry into a familiar surface. It made Einstein's idea feel understandable. Suddenly, people could imagine gravity not as a mysterious pull, but as curved spacetime. The Sun bent the fabric. Earth followed the curve. Planets moved along the geometry. Light bent because the path itself curved.

For teaching, this was useful.

But usefulness has a danger.

The metaphor slowly became more powerful than the caution attached to it.

A rubber sheet is not spacetime. It is only an analogy. It is a two-dimensional visual aid used to represent something far more complex. But once the image enters the mind, it begins to behave like reality. People start imagining spacetime as a literal cosmic cloth. They imagine stars sitting on it, planets rolling around depressions, black holes making deep pits, and gravitational waves shaking the fabric.

The teaching picture becomes the physical picture.

This is where confusion begins.

If spacetime is a fabric, what is the fabric made of?

If it bends, what exactly is bending?

If it stretches, what is being stretched?

If it ripples, what carries the ripple?

If it guides motion, what gives it this physical capacity?

These questions are often hidden behind the elegance of the metaphor. The mind becomes satisfied too early because the picture looks convincing. But a convincing image is not the same as a complete mechanism.

The fabric metaphor also creates another issue. In the rubber-sheet picture, the sheet bends because gravity is already pulling the heavy ball downward. But if gravity is what we are trying to explain, then using gravity to explain the bending of the sheet creates circular imagination. The analogy works only because we already live inside gravity. A ball depresses a rubber sheet because Earth's gravity pulls it down. But in space, there is no larger external gravity pulling the Sun "down" into spacetime fabric.

So the analogy is helpful, but incomplete.

It explains the result visually, not the deeper physical cause.

This does not weaken Einstein's mathematical achievement. Einstein did not merely invent a rubber-sheet story. His framework is far more rigorous than the public metaphor. The equations describe relationships between energy, momentum, curvature, and motion. They have been tested successfully in many domains. Their predictive power cannot be dismissed casually.

But the philosophical question remains.

What is the physical nature of the thing being geometrically described?

If spacetime is only a mathematical structure, then curvature is a description of relational behavior, not necessarily a material bending. If spacetime is treated as physically active, then it cannot be pure emptiness. Something that curves, responds, transmits waves, and determines paths cannot be absolute nothingness.

This is where Universal Energy Dynamics asks for a deeper interpretation.

UED does not reject the observed behavior that general relativity describes. It questions whether "spacetime fabric" is the best physical language for explaining that behavior. Perhaps what is being described as curvature is actually a deeper field response. Perhaps the so-called fabric is not spacetime itself, but Structured Vacuum Energy behaving through gradients, resistance, and equilibrium-seeking dynamics.

The difference is subtle but important.

In the spacetime-fabric imagination, mass bends the stage.

In UED, mass is itself a field signature, and gravity emerges from the energy-gradient behavior of the field.

This removes the need to imagine empty spacetime as a magical cloth. The medium is no longer nothing pretending to bend. The medium is Structured Vacuum Energy, the deeper field condition from which matter, mass, motion, and gravitational behavior emerge together.

The human imagination turned gravity into fabric because it needed a picture.

That picture helped.

But now the picture may be limiting us.

A metaphor is a bridge, not a home. It helps us cross from ignorance toward understanding. But if we build our permanent house on the bridge, we confuse movement with destination. The fabric metaphor was a bridge. It helped many minds escape the old image of gravity as invisible pulling. But perhaps now we must cross further.

Gravity may not be fabric bending.

It may be field imbalance resolving.

It may not be geometry alone.

It may be energy-gradient consequence.

The universe may not be a cloth stretched across nothing.

It may be a structured energy field revealing curvature-like behavior when equilibrium is disturbed.

5.2 — The Problem with Empty Spacetime

The word "empty" creates many problems.

It looks simple. It feels clean. It gives the mind a blank stage on which objects can move. Matter here, space there. Stars here, emptiness between them. Galaxies here, vast nothingness separating them. This picture is easy to imagine because daily experience supports it. A room can be empty. A box can be empty. A road can be empty. So we extend the same habit to the cosmos and imagine space as emptiness.

But the universe does not behave like an empty box.

This is the first difficulty.

If spacetime is truly empty, then how does it curve? How does it ripple? How does it guide motion? How does it carry gravitational waves? How does it respond to mass and energy? How does it allow light to bend, clocks to slow, or trajectories to change?

Something that behaves cannot be absolute nothingness.

This is the philosophical weakness hidden inside the phrase "empty spacetime." On one side, spacetime is described as a geometric structure capable of curvature. On the other side, it is often imagined as empty background. But emptiness cannot possess behavior. Absolute nothing cannot bend, stretch, transmit, resist, guide, or structure anything.

If it acts, it is not nothing.

This does not mean spacetime must be a material fabric like cloth. That would be another mistake. But it does mean that our language is incomplete. We cannot keep saying that empty spacetime bends unless we clarify what kind of physical reality is being described by that bending.

The problem becomes clearer when we ask a simple question.

If there were no matter, no energy, no field, no disturbance, no relation, no resistance, no interaction, and no possibility of consequence, what would "spacetime" mean physically?

Would it still exist as a real thing?

Would it still curve?

Would it still carry waves?

Would it still produce measurable effects?

These questions are not childish. They are foundational.

Modern physics often survives by mathematical description. If the equations work, the model gains authority. That is fair within science. But equations describe relationships. They do not always settle the deeper physical meaning of the relationships they describe. A mathematical geometry may successfully model gravitational behavior, but the philosophical question remains: what is the underlying reality that behaves geometrically?

Universal Energy Dynamics answers this differently.

It does not begin with empty spacetime.

It begins with Structured Vacuum Energy.

In UED, the vacuum is not dead absence. It is the structured field condition from which matter, mass, gravity, light, and observable consequences emerge. It is not empty background. It is the deeper medium of possibility, resistance, ripple behavior, and equilibrium-seeking response.

This changes the meaning of gravity.

If the vacuum is structured, then gravitational behavior does not require an empty fabric to bend mysteriously. The field itself responds to energy condensation. Where energy density changes, gradients appear. Where gradients appear, imbalance seeks resolution. What we call gravity is the observable behavior of this field trying to restore balance.

The bending-like behavior is not denied.

It is reinterpreted.

Instead of saying that empty spacetime curves around mass, UED says that energy condensation within Structured Vacuum Energy creates a field-gradient condition. Motion, orbit, clustering, and bending are consequences of this gradient. The geometry may describe the behavior, but the physical process belongs to field imbalance and equilibrium.

This is an important distinction.

Geometry tells us how motion appears.

Field dynamics tells us why the condition exists.

If we stop at geometry, we may calculate accurately but still misunderstand the underlying mechanism. If we go deeper into field dynamics, we begin to understand gravity not as a mysterious property of spacetime, but as an expression of structured energy behavior.

This also connects directly to Chapter 4.

Matter is not an independent thing. Matter is ripple condensation within Structured Vacuum Energy. Mass is not a fixed possession. Mass is field-density expression. Therefore, gravity cannot be something that begins only after matter appears and then acts upon it from outside. Matter, mass, and gravity must emerge together from the same deeper field process.

Where the field condenses, mass becomes measurable.

Where mass becomes measurable, gradients become visible.

Where gradients exist, motion follows the path of imbalance.

This is gravity under UED.

The old image says: matter bends spacetime.

The UED image says: energy condensation creates field gradients within Structured Vacuum Energy, and those gradients produce gravitational behavior.

The first depends on a fabric metaphor.

The second depends on field consequence.

This removes the awkwardness of empty spacetime. We no longer need to imagine nothing behaving like something. We no longer need to treat emptiness as both absence and active structure. The vacuum becomes meaningful again. It becomes the silent participant in all observable phenomena.

This also helps us understand why gravity is universal.

If gravity were merely a force between objects, it would remain tied to the object-based imagination. If gravity were merely curvature of an abstract geometry, it would remain difficult to connect with physical mechanism. But if gravity is energy-gradient behavior within the structured field, then it naturally appears wherever energy condensation, resistance, and imbalance exist.

Gravity is not added to the universe.

It emerges from the way the universe seeks balance.

This view also prevents a common misunderstanding. Saying that spacetime fabric is a myth does not mean gravitational observations are false. Objects still fall. Planets still orbit. Light still bends. Clusters still form. Gravitational waves may still be detected as measurable disturbances. The question is not whether these effects occur. The question is what physical language best explains them.

UED proposes that the language of empty spacetime is too thin.

It lacks the medium.

It describes the curve but forgets the condition.

It gives us geometry but not enough mechanism.

Structured Vacuum Energy gives gravity a deeper home. It allows us to see gravitational behavior as part of the same field reality that produces matter and mass. It unifies what the object-based model separates. Matter, gravity, motion, structure, and time become related consequences of energy behavior, not isolated categories forced to interact from outside.

This matters greatly for cosmology.

If spacetime is treated as empty fabric, then galaxies are structures bending that fabric according to their mass. If the observed motion does not match visible mass, we must add unseen mass or modify the equations. But if the vacuum itself is structured and gravity is energy-gradient behavior, then galactic motion may reflect deeper field conditions not captured by visible matter alone.

This does not automatically solve every problem.

But it changes the direction of inquiry.

Instead of asking only, "What hidden matter is bending spacetime?" we may ask, "What field-gradient condition is producing this gravitational behavior?"

Instead of asking only, "How does mass curve emptiness?" we may ask, "How does structured energy create the gradients we observe as gravity?"

These questions open a wider path.

Empty spacetime may have been a useful abstraction. But reality may not be abstract emptiness. It may be active field condition. It may be Structured Vacuum Energy expressing imbalance, resistance, ripple, and balance.

If something bends, something responds.

If something carries consequence, it is not nothing.

And if gravity is the consequence we observe, then perhaps the universe is not bending empty space.

Perhaps it is revealing the hidden structure of the vacuum itself.

5.3 — Mass Is Not the Cause Alone; It Is a Field Signature

In the usual understanding of gravity, mass is treated as the great cause.

Where there is mass, there is gravity. More mass means stronger gravity. A planet pulls a stone. A star holds planets. A galaxy governs stars. A black hole dominates everything near it because its mass is extreme. This way of thinking is familiar, useful, and mathematically workable across many situations.

But Universal Energy Dynamics asks us to look one layer deeper.

What is mass?

If mass is treated as an independent possession of matter, then gravity naturally appears as something caused by that possession. Matter has mass, mass creates gravity, gravity controls motion. The explanation feels complete because each step follows the previous one.

But the foundation remains unexamined.

If matter itself is not an independent substance, but ripple condensation within Structured Vacuum Energy, then mass cannot be understood as an isolated property sitting inside matter. Mass becomes a measurable signature of the field condition that has condensed into matter-form. It is not the beginning of gravity. It is one of the visible signs that the field has reached a certain density, coherence, resistance, and stability.

This changes the entire sequence.

The old view says: matter exists, matter has mass, mass produces gravity.

UED says: structured energy condenses, mass becomes measurable, field gradients emerge, and gravitational behavior appears as the field seeks balance.

Mass is no longer the master cause.

Mass is a field signature.

This does not deny the relationship between mass and gravity. It deepens it. A massive object still shows stronger gravitational influence. A star still shapes planetary motion. A galaxy still carries large-scale gravitational behavior. But the reason is not simply that mass commands the universe like an external authority. The reason is that mass reveals a deeper condition of energy condensation within the field.

Where energy is densely condensed, the surrounding field is not neutral.

It carries a gradient.

That gradient is the beginning of gravitational behavior.

This is why the phrase "mass causes gravity" may be too simple. It works at the surface level, but it hides the underlying field process. It is like saying clouds cause rain without asking how clouds form. It is like saying a whirlpool pulls floating leaves without asking what flow condition created the whirlpool. It is like saying a flame gives heat without asking what process sustains the flame.

The visible cause may not be the deepest cause.

Mass is visible through measurement. Field condition is inferred through deeper interpretation. Because mass is easier to measure than the field condition behind it, science naturally gives mass primary status. But ease of measurement should not be confused with final reality.

In UED, mass is the measurable density of stabilized energy.

It is how the field announces that a local condensation has become coherent enough to resist disturbance and interact with surrounding field conditions. The more intense and stable the condensation, the stronger the relational gradient. The stronger the gradient, the stronger the gravitational behavior.

Gravity therefore does not begin as a mysterious pull between isolated masses.

It begins as an imbalance relationship inside the field.

This point is crucial because it removes the false separation between matter and gravity. In the old imagination, matter appears first, then gravity acts between matter. In UED, matter and gravity are connected consequences of the same energy-field behavior. Matter is condensed structure. Mass is measurable field density. Gravity is the gradient response around that density.

One process, many expressions.

This also helps us understand why gravity appears universal. It is not universal because every object secretly carries an invisible rope. It is universal because every condensed field structure alters the equilibrium condition around it. Wherever energy condenses into stable form, the surrounding field must respond. The response appears as gravitational tendency.

A planet is not pulling a stone in the crude sense of a hand pulling an object.

The planet represents a massive field condensation. The stone represents another condensed field structure. Between them, and around them, the Structured Vacuum Energy carries gradient relations. Motion occurs according to the way the field seeks balance between these conditions.

This is why gravity is relational.

No mass exists in total isolation. No object is completely separate from the field. Every form is a local expression within a continuous field. Its gravitational behavior is therefore not merely its private property. It is the relational consequence of its field presence.

This has direct implications for cosmology.

If mass is treated only as material inventory, then gravitational behavior must be explained by counting visible matter and adding invisible matter where the count fails. This is how modern cosmology becomes dependent on dark matter in many contexts. The visible mass does not explain the observed motion, so unseen mass is introduced as additional gravitational cause.

UED does not casually dismiss such inference. It simply asks whether the starting interpretation is complete.

What if the gravitational behavior is not caused only by hidden material mass?

What if it reflects deeper field-gradient conditions that are not captured by visible matter alone?

What if a galaxy's gravitational identity is not merely the sum of its stars, gas, dust, and inferred halos, but the expression of a large-scale field configuration?

Then the search for missing mass may be partly a search created by treating mass too materially.

The field may be doing what invisible matter was asked to do.

This is not an escape from explanation. It is a demand for deeper explanation. If gravity is energy-gradient behavior, then we must study the field condition itself, not only the material inventory. A galaxy may remain coherent not simply because enough hidden stuff is holding it together, but because its entire structure is a gradient-stabilized expression within SVE.

This also changes the meaning of black holes.

If mass is field signature, then extreme gravitational behavior does not automatically require imagining an infinite-density object. It may indicate an extreme field condition where energy condensation, gradient intensity, and coherence collapse reach a saturation boundary. What mainstream science calls a black hole may be less a thing of infinite mass-density and more a zone where field behavior exceeds ordinary observational recovery.

That discussion will come later.

For now, the important point is this: mass should not be worshipped as the final cause.

Mass is not the king sitting above the field.

Mass is the field showing local density.

Once this is understood, gravity becomes less like a command and more like a consequence. It is not an independent force imposed upon matter. It is the behavior that appears when condensed energy creates imbalance inside a continuous field.

The universe does not say, "Mass exists; therefore everything must pull."

The universe says, "Where energy condenses, balance is disturbed; where balance is disturbed, gradients arise; where gradients arise, motion follows."

That motion is what we call gravity.

This is why the UED reinterpretation is necessary. It does not deny that mass and gravity are related. It refuses to stop at the surface of that relation. It asks what mass itself means, where it comes from, and how it connects to the field.

Mass is not the cause alone.

Mass is the clue.

The deeper cause is structured energy seeking equilibrium through the field.

5.4 — Gravity as Energy-Gradient Equilibrium

Gravity may not be a pull.

It may be balance trying to restore itself.

This is the central shift of Universal Energy Dynamics.

In the ordinary imagination, gravity is understood as attraction. One object pulls another. Earth pulls the stone. The Sun pulls the planets. A galaxy pulls stars into orbit. The stronger the mass, the stronger the pull.

This language is convenient because it matches the way motion appears from the outside.

But appearance is not always mechanism.

When a stone falls toward Earth, we say Earth pulls it. But what if the deeper process is not pulling at all? What if the stone is moving along a field-gradient condition created by energy imbalance? What if motion occurs not because one object commands another, but because the field is seeking equilibrium between different energy-density states?

This is the UED view.

Gravity is the emergent consequence of energy gradients seeking balance within Structured Vacuum Energy.

Where energy is condensed, the field condition changes. Where the field condition changes, a gradient appears. Where a gradient appears, the system develops directional tendency. What we observe as falling, orbiting, clustering, bending, and structural coherence may be the visible expression of this tendency.

In this interpretation, gravity is not an independent force added to reality.

It is the behavior of imbalance.

This is important because imbalance is not a defect. Imbalance is the beginning of action. A perfectly balanced field would not need motion. It would not need correction. It would not need redistribution. But once energy density differs from one region to another, the field cannot remain neutral. It begins to respond. That response appears as motion toward balance.

Gravity is that response seen at the cosmic scale.

A planet is not merely a ball of matter pulling nearby objects. It is a condensed energy configuration that creates a surrounding field-gradient condition. A falling stone is not merely a passive object being dragged downward. It is another condensed structure moving according to the relational gradient between itself and Earth's field condition.

The motion looks like attraction.

The deeper mechanism may be equilibrium-seeking.

This distinction changes the emotional image of gravity. In the old language, gravity is almost aggressive. It pulls, captures, collapses, traps, and dominates. Massive bodies seem to rule smaller bodies. Black holes become monsters. Galaxies become gravitational prisons. Everything appears to be controlled by invisible force.

But in UED, gravity is not domination.

Gravity is correction.

It is the field's way of reducing imbalance. It is structured energy attempting to resolve difference. It is the universe moving toward a more stable configuration. What looks like attraction may be the field's natural effort to contract delta between energy states.

This is where the phrase "energy-gradient equilibrium" becomes essential.

A gradient means difference. One region has a different energy condition than another. This difference may involve density, condensation, resistance, coherence, or field tension. Because the field is structured, the difference does not remain meaningless. It creates directional behavior. The system begins moving toward equilibrium, not because it has intention like a human mind, but because structured energy cannot remain indifferent to imbalance.

The field seeks balance because balance is the natural tendency of structured reality.

In this sense, gravity is not the beginning of structure.

Gravity is the consequence of structure already disturbing balance.

When energy condenses into matter, the surrounding field must adjust. The condensation creates a local difference. That difference becomes a gradient. The gradient influences motion. The motion creates measurable gravitational behavior. Therefore, gravity is not separate from matter, mass, or field. It is their relational consequence.

This also helps explain why gravity behaves across distance.

In object-based thinking, action across distance appears mysterious. How does one body affect another across emptiness? Newton gave the mathematics but did not fully explain the mechanism. Einstein replaced action-at-a-distance with spacetime curvature, but the physical nature of the curving structure remains philosophically difficult if spacetime is imagined as empty.

UED removes this gap by refusing emptiness.

There is no dead nothing between objects.

There is Structured Vacuum Energy.

The field is continuous. Therefore, energy gradients are not isolated local events. They are relations within the continuous field. A planet, star, or galaxy does not influence surroundings through magic. Its condensed energy state modifies the field condition around it. Other condensed structures move according to the gradient relationships within that same field.

This makes gravity relational, not mechanical in the crude sense.

It is not one object throwing invisible hooks at another.

It is the continuous field expressing directional behavior where balance is incomplete.

A useful analogy is a slope. A ball rolls downhill not because the bottom of the hill pulls it with a rope, but because the surface condition creates a direction of motion. The ball follows the gradient. Of course, the analogy is limited because a hill exists under gravity, so it cannot explain gravity itself. But it helps us understand the difference between pulling and gradient-following.

Gravity may be closer to gradient-following than pulling.

Another analogy is pressure difference. Air moves from high-pressure regions to low-pressure regions. We may say wind blows, but wind is not commanded by a single object. It is the visible movement created by differences in pressure. The difference produces direction. The direction produces motion. The motion becomes experience.

Similarly, gravity may be the cosmic behavior produced by differences in energy condition.

Where energy density is high, the field curves—not necessarily as a cloth, but as a condition of relational resistance and balance. Where energy density is lower, the field condition differs. The gradient between these states guides motion. What we call gravitational attraction is the observable result.

This also changes the meaning of orbit.

An orbit is not simply a smaller body being trapped by a larger body. It is a dynamic balance between motion and field-gradient response. The planet continues moving, but its path is shaped by the gradient condition created by the star's field density. The orbit is not a prison. It is a sustained negotiation between energy states.

This is why cosmic structure can remain stable.

Galaxies, solar systems, clusters, and planetary systems are not merely collections of objects held together by invisible force. They are field-gradient arrangements. Their stability comes from the ongoing balance of motion, condensation, resistance, and gradient behavior. When the balance shifts, structure changes.

This is extremely important for interpreting JWST observations.

If gravity is understood only as the slow organizer of collected matter, then galaxies must take long time to form. Matter must gather gradually under gravitational attraction. Large structures must grow step by step. Mature galaxies at cosmic depth become difficult.

But if gravity is energy-gradient equilibrium, then structure may arise more directly from field conditions. Where energy condensation creates strong gradients, field realignment can produce coherent structure. Gravity is not waiting for matter to assemble slowly; gravity emerges with the condensation and participates in the stabilization of structure.

The galaxy is not first a pile of matter and then a gravitational system.

It is a field-gradient system from the beginning of its manifestation.

This is why UED can reinterpret compact high-redshift galaxies and Little Red Dots differently. Their compactness may indicate strong local energy condensation. Their brightness may indicate intense field realignment. Their apparent maturity may indicate coherence of the gradient structure. They are not necessarily objects that gravity assembled too quickly. They may be field-gradient consequences that became visible when the conditions crossed threshold.

This also reduces dependence on invisible scaffolding.

If gravity is treated as only the effect of material mass, then missing gravitational behavior must be explained by missing matter. But if gravity reflects deeper field-gradient conditions, then the observed motion of galaxies may reveal field structure not captured by visible matter alone. The missing factor may not always be unseen material. It may be misunderstood field behavior.

This must be handled carefully.

UED does not casually say that every existing dark matter inference is meaningless. It says the interpretation may be incomplete if the field itself is not properly included. The gravitational behavior attributed to hidden mass may partly arise from gradient logic within Structured Vacuum Energy.

This makes gravity less isolated and more universal.

It connects gravity to matter, mass, vacuum, redshift, structure, and time. It shows that gravity is not a separate mystery floating beside other mysteries. It is one expression of the same deeper principle: structured energy seeking equilibrium through consequences.

The universe does not pull for the sake of pulling.

It balances.

It does not curve emptiness for the sake of geometry.

It reveals gradients.

It does not create gravity as an independent force imposed from outside.

It expresses gravity wherever energy condensation disturbs equilibrium.

This is why the myth of spacetime fabric must be reconsidered. The fabric metaphor may describe the appearance of curved paths, but UED asks what the field is doing beneath that appearance. Gravity is not merely the shape of spacetime. It is the behavior of structured energy when balance is incomplete.

At the deepest level, gravity may be the universe saying:

Here, equilibrium is not finished.

Here, energy density differs.

Here, the field must respond.

Here, motion will follow the gradient until balance is approached.

This is not attraction as command.

This is balance as consequence.

5.5 — Why Galaxies Do Not Need Invisible Scaffolding First

Modern cosmology often explains galaxies as structures that require invisible scaffolding.

The visible matter alone does not seem enough. Stars rotate in ways that do not comfortably match the gravitational influence of the matter we can directly observe. Galaxies appear more stable than expected. Clusters behave as if additional mass is present. Large-scale structure seems to require an unseen framework. So the standard explanation introduces dark matter as the hidden support system beneath visible cosmic architecture.

This idea has become central to modern cosmology.

In the standard story, dark matter does not shine, does not emit light, does not interact electromagnetically in ordinary ways, but produces gravitational effects. It becomes the invisible skeleton around which galaxies form. Ordinary matter falls into dark matter halos. Gas settles. Stars form. Galaxies grow inside these unseen gravitational wells.

This is the scaffolding model.

It is not a foolish idea. It arises from real observational difficulties. When visible matter does not explain motion, structure, and stability, science must search for what is missing. The dark matter hypothesis is one attempt to account for that missing gravitational behavior.

But Universal Energy Dynamics asks whether the missing element has been interpreted correctly.

What if the problem is not simply missing matter?

What if the problem is a matter-first understanding of gravity?

If gravity is assumed to be caused only by material mass, then any unexplained gravitational behavior naturally demands more mass. If visible mass is insufficient, invisible mass must be added. The logic is straightforward. But it depends on the assumption that gravity is primarily a mass-produced effect.

UED changes that assumption.

If gravity is energy-gradient behavior within Structured Vacuum Energy, then gravitational behavior does not need to be explained only by counting visible and invisible matter. It must also be explained by understanding the field-gradient condition of the region itself.

This is a major shift.

The galaxy may not be held together merely by an invisible halo of extra material. It may be stabilized by the gradient logic of the field. Its visible stars, gas, dust, central regulator, motion, and surrounding field condition may all belong to one coherent energy arrangement within Structured Vacuum Energy.

In this view, a galaxy is not a crowd of objects needing hidden glue.

It is a field configuration.

This does not mean there is no unseen component in the universe. It does not mean all dark matter discussions are worthless. It means the interpretation of missing gravity may be incomplete if the field itself is ignored. The unseen gravitational behavior may not always indicate unseen matter. It may indicate unseen field structure.

That distinction is crucial.

Matter-first cosmology sees a galaxy and asks: how much mass is needed to produce this motion?

UED sees a galaxy and asks: what field-gradient condition produces this motion?

The first question searches for inventory.

The second searches for mechanism.

This does not remove measurement. It deepens it. Rotation curves, lensing behavior, cluster dynamics, and galactic stability remain important. But instead of immediately translating every unexplained motion into missing matter, UED asks whether Structured Vacuum Energy is carrying gradient behavior that visible matter alone cannot represent.

A galaxy may be like a whirlpool.

The visible motion of a whirlpool cannot be explained by counting only the foam on the surface. The foam reveals the motion, but it does not produce the entire system. The deeper water flow, pressure differences, boundaries, and rotational dynamics sustain the structure. If someone saw only the foam and tried to explain the whirlpool by adding invisible foam, the explanation would miss the medium.

Similarly, visible stars may be the foam of the galactic field.

They reveal the structure, but they may not fully explain the field condition sustaining it.

The galaxy's coherence may arise from a deeper SVE gradient pattern. The visible matter is part of that pattern, not the complete cause of it. The gravitational behavior belongs to the whole field arrangement, not merely to the visible material inventory.

This is why dark matter scaffolding may be a patch over a deeper misunderstanding.

The standard model needs invisible scaffolding because it treats galaxies as material objects requiring gravitational support. But if galaxies are field consequences from the beginning, then their stability does not need to be imagined as something added from outside. Stability belongs to the field structure itself.

The galaxy is not first built and then supported.

It is expressed as a supported field condition.

This changes the meaning of formation. In the scaffolding model, dark matter halos form first, and ordinary matter later gathers inside them. Under UED, the galaxy arises through field realignment, ripple condensation, and energy-gradient stabilization. What appears as gravitational support may be the field's own equilibrium structure, not a separate invisible substance surrounding the galaxy.

This matters especially for early galaxies and Little Red Dots.

If mature structures appear at cosmic depth, the standard model must explain how dark matter halos, gas collapse, star formation, and galactic organization happened quickly enough. But if gravitational coherence is field-gradient behavior, then compact structure may emerge wherever the field condition reaches alignment threshold. The galaxy does not need to wait for invisible scaffolding to assemble first. The field condition itself is the scaffolding.

This is the UED proposition.

The scaffold is not hidden matter.

The scaffold is structured vacuum behavior.

This also changes how we read galactic rotation. In the standard view, if outer stars move faster than expected, additional mass must be present. In UED, the motion may reflect the larger field-gradient geometry of the galaxy. The stars are not orbiting merely under the pull of visible mass or hidden material. They are moving within a structured gradient condition created by the galaxy as a whole field system.

The galaxy is not a pile.

It is a pattern.

A pile needs external support.

A pattern needs sustaining conditions.

The same difference applies across cosmology. If the universe is viewed as a collection of objects, unseen matter becomes necessary wherever visible objects fail to explain behavior. But if the universe is viewed as a structured field, then behavior may arise from conditions that are not material in the ordinary sense.

This is why UED does not rush to fill every gap with another substance.

It asks whether the gap exists because the field was excluded from the explanation.

A galaxy may not need invisible scaffolding first because it is not formed as a dead material structure requiring support afterward. It is born as a field-gradient configuration. Its matter, motion, brightness, central regulation, and gravitational behavior are connected consequences of the same underlying energy arrangement.

This does not make the problem simpler in a lazy way.

It makes it deeper.

Instead of saying, "Something invisible must be there because the motion demands it," UED asks, "What is the field doing that makes this motion appear necessary?" Instead of treating gravity as evidence of hidden material only, UED treats gravity as evidence of imbalance, gradient, and field coherence.

This is a more integrated view.

It does not separate matter from vacuum, mass from field, gravity from energy, or structure from equilibrium. It sees galaxy formation as a field event. It sees galactic stability as a consequence of sustained gradient balance. It sees cosmic motion as the visible behavior of structured energy trying to resolve itself.

If this interpretation is correct, then one of the biggest assumptions of modern cosmology must be re-examined.

Perhaps galaxies are not held together by invisible extra matter alone.

Perhaps they are held together by the same field logic that expressed them.

The universe may not need hidden scaffolding placed behind visible structure.

The structure itself may be the visible face of the scaffold.

5.6 — Gravity, JWST, and the End of Slow Assembly

JWST did not merely challenge the age of galaxies.

It challenged the speed of structure.

In the standard interpretation, galaxies are expected to form gradually. Matter gathers. Gas cools. Stars ignite. Smaller structures merge. Dark matter halos provide gravitational scaffolding. Over long periods, the galaxy becomes larger, brighter, more organized, and more mature. This is the slow assembly model, and gravity plays the role of the patient builder.

Gravity pulls.

Matter gathers.

Structure grows.

Time completes the work.

This sounds reasonable within the old framework. If matter is treated as independent substance and gravity is treated as the force or curvature organizing that substance, then cosmic structure must require duration. A mature galaxy becomes the outcome of a long sequence. It cannot appear too early because the builder has not had enough time.

But JWST disturbed this expectation.

It revealed deep-field structures that do not always look like weak beginnings. Some appear bright, compact, dense, and organized. Some appear as if the universe had already achieved structural coherence where the timeline expected early formation. The discomfort is not merely observational. It is mechanical. The old story asks: how did gravity assemble so much, so fast?

This question already carries an assumption.

It assumes that gravity assembles structure from matter over time.

Universal Energy Dynamics changes that assumption.

If matter is ripple condensation within Structured Vacuum Energy, and if gravity is energy-gradient behavior seeking equilibrium, then galaxies are not merely objects slowly assembled after matter appears. They are field-gradient systems from the beginning of their manifestation. Matter, mass, gravitational behavior, brightness, compactness, and structure are connected consequences of one deeper field condition.

This changes the meaning of formation.

In the old model, gravity comes after matter. Matter appears, mass exists, gravity acts, and structure forms. In UED, the sequence is not so separated. Structured energy condenses. Mass becomes measurable. Field gradients appear. Gravitational behavior emerges. Structure stabilizes. These are not independent steps placed far apart in cosmic time. They are interconnected expressions of the same field process.

The galaxy is not first a pile of matter and later a gravitational structure.

It is born as a gradient condition.

This is why mature-looking deep galaxies may not require the same slow assembly that mainstream cosmology expects. Their structure may not be the final product of long material accumulation alone. It may be the visible expression of field realignment, ripple condensation, and gradient stabilization within Structured Vacuum Energy.

In that case, gravity is not merely the slow laborer carrying matter into place.

Gravity is the sign that field imbalance is already active.

Where energy density differs, gradients arise. Where gradients arise, motion becomes directional. Where motion and condensation stabilize together, structure appears. The galaxy is therefore not only gathered; it is expressed. Its gravitational behavior is not added afterward; it belongs to its field identity.

This is especially important for Little Red Dots and compact high-redshift structures.

If they are interpreted through slow assembly, they become puzzles. How did they become so compact? How did they shine so strongly? How did so much apparent mass or energy concentrate within the assigned early period? Why do they look structurally intense when the timeline expected primitive looseness?

But through UED, compactness may indicate strong local field condensation. Brightness may indicate energetic release from ripple alignment. Redness may carry not only distance, but field resistance and signal-condition information. Apparent maturity may indicate coherence, not old age.

The object is no longer accused of forming too fast.

It is read as a field condition that has reached threshold.

This is the end of slow assembly as the only explanation.

It does not mean galaxies never evolve. They do. They may interact, merge, redistribute, stabilize, transform, and dissolve. Their internal structures may change over measurable durations. Stars may form and die. Regions may brighten or fade. Central regulators may influence the whole system. None of this is denied.

But the foundational appearance of structure may not depend only on slow mechanical gathering.

It may depend on field readiness.

This is a decisive shift.

A whirlpool may develop quickly when the flow condition is right. A flame may appear when ignition conditions are met. A crystal may form when the medium reaches the right state. These examples are not perfect cosmic equivalents, but they reveal a principle: structure can emerge rapidly when the medium already contains the conditions for organized expression.

The same may be true at cosmic scale.

A region of Structured Vacuum Energy may carry intense gradients, resistance, and ripple tension. When the condition crosses the manifestation threshold, matter condenses, mass becomes measurable, gravity-like behavior emerges, and structure appears as a coherent field consequence. To a timeline-based model, this looks too fast. To a field-based model, it is threshold behavior.

JWST may be showing us exactly this.

Not galaxies that broke the rules.

Not structures that violated time.

Not cosmic objects that accidentally matured too early.

But field-gradient systems that became visible through a telescope powerful enough to resolve deeper layers of cosmic structure.

This also weakens the need to treat gravity as a simple assembly tool. Gravity, under UED, is not only the mechanism by which finished matter gathers. It is the emergent behavior of energy imbalance while matter itself is forming as condensation. Gravity and matter are not separate actors meeting later on the stage. They are co-expressions of structured field dynamics.

This explains why the old question may be too limited.

"How did gravity build these galaxies so quickly?"

The deeper question is:

"What field-gradient condition expressed these galaxies so coherently?"

The first question assumes slow construction.

The second opens the door to field manifestation.

This does not make the universe less scientific. It makes the science more fundamental. Instead of forcing every structure into a construction timeline, UED asks what kind of field condition made the structure possible. Instead of treating gravity as an invisible worker arranging matter, it treats gravity as the observable signature of imbalance seeking resolution.

Then JWST becomes less of a crisis-maker and more of a revealer.

It reveals that cosmic structure may be faster than mechanical assembly allows because structure may not be only mechanical assembly. It reveals that depth may contain coherence because coherence depends on field condition, not just chronological age. It reveals that galaxies may appear mature not because they cheated time, but because time was never the only builder.

Gravity, in this new reading, does not wait patiently for scattered matter to slowly organize.

Gravity emerges wherever structured energy creates imbalance.

And where imbalance seeks balance, structure can appear.

This is why JWST matters so deeply to the UED reinterpretation. It does not merely give us sharper images. It forces us to reconsider the mechanism behind cosmic formation. If galaxies at depth appear too coherent for slow assembly, perhaps the answer is not only faster assembly. Perhaps the answer is that assembly itself is not the deepest mechanism.

The universe may not be building galaxies like a mason placing stones.

It may be aligning them like a field resolving tension.

And once gravity is understood as energy-gradient equilibrium, the old cosmic construction story loses its monopoly. Structure no longer depends only on collected matter and accumulated time. It depends on field condition, condensation, resistance, gradient, and balance.

JWST did not show us that gravity worked too fast.

It may have shown us that gravity was never merely a slow builder.

It was always the field revealing imbalance and moving toward coherence.

Closing Reflection — Gravity Does Not Pull; It Reveals Imbalance

Gravity does not need to be imagined as an invisible hand.

It does not need to be treated as a mysterious rope stretched between objects. It does not need to be locked forever inside the image of a fabric bending beneath mass. These images helped us think, but perhaps they also trained us to stop thinking too early.

A falling stone is real.

An orbiting planet is real.

A rotating galaxy is real.

The bending of light near massive structures is real.

The question is not whether gravitational behavior exists. The question is whether our explanation of it has reached the deepest possible level.

For centuries, gravity has been described through the language of attraction. One body pulls another. Later, this language evolved into geometry. Matter curves spacetime, and objects follow the curvature. Both explanations gave humanity extraordinary power. Newton gave calculation. Einstein gave elegance. Together, they shaped the scientific imagination of the cosmos.

But Universal Energy Dynamics asks us to go beneath both.

If matter is not an independent substance but ripple condensation within Structured Vacuum Energy, then mass is not merely a private property carried by objects. Mass is a field signature. It is the measurable presence of condensed energy. And if mass itself is a field signature, then gravity cannot be understood merely as something that mass "creates" from outside the field.

Gravity must be a field consequence.

This is the central landing of this chapter.

Gravity is the observable behavior of energy gradients seeking balance. Wherever energy condenses, the field condition changes. Wherever the field condition changes, imbalance appears. Wherever imbalance appears, directional tendency arises. That directional tendency becomes visible as falling, orbiting, clustering, rotating, bending, and structural coherence.

In ordinary language, we call it gravity.

But in deeper language, it is equilibrium trying to complete itself.

This changes the emotional meaning of gravity.

Gravity is often imagined as domination. Massive objects control smaller ones. Stars hold planets. Galaxies hold stars. Black holes trap light. The universe seems governed by invisible capture. But in UED, gravity is not domination. It is correction. It is not the universe forcing objects into obedience. It is the field responding to imbalance.

Gravity reveals where balance is unfinished.

This is why the idea of empty spacetime fabric becomes insufficient. If something bends, responds, ripples, and guides motion, it cannot be pure nothingness. The vacuum cannot be dismissed as absence. It must be understood as active condition. Under UED, that condition is Structured Vacuum Energy, the silent medium through which matter appears, mass becomes measurable, and gravity emerges as gradient behavior.

The old model says: mass bends spacetime.

UED says: energy condensation creates field gradients within Structured Vacuum Energy.

The old model says: gravity organizes matter.

UED says: matter and gravity arise together as expressions of field imbalance and field resolution.

The old model says: galaxies need hidden scaffolding when visible matter is insufficient.

UED asks: what field-gradient structure is being misread as missing material?

This does not mean every established gravitational calculation becomes useless. It means calculation and interpretation are not the same. A mathematical model may describe behavior accurately within a certain frame, but the physical meaning of that behavior may still require deeper revision. To calculate the path of a stone is one thing. To understand why the field permits that path is another.

The same applies to galaxies.

If gravity is treated as a slow builder acting upon matter, then mature-looking galaxies at cosmic depth become difficult. There seems to be too little time for enough matter to gather and organize. But if gravity is energy-gradient equilibrium, then galactic structure is not merely accumulated matter shaped by delayed attraction. It is a field-gradient configuration becoming visible when conditions reach threshold.

Then JWST's disturbing observations begin to look different.

Compact galaxies, bright deep structures, and Little Red Dots may not be objects that gravity assembled too quickly. They may be field consequences whose gravitational coherence appeared with the very act of manifestation. Their structure may not be a violation of time. It may be the signature of field alignment.

This is the deeper resolution.

Gravity is not the universe pulling things together for no reason.

Gravity is the universe revealing where energy distribution has not yet settled into balance.

A galaxy rotates because the field is not empty.

A planet orbits because the star is not merely an object, but a field-density center.

A stone falls because Earth is not merely a mass below it, but a gradient condition within Structured Vacuum Energy.

A black hole candidate becomes extreme not because reality contains an infinite monster, but because field saturation may have reached a point where ordinary coherence collapses from observation.

Everything changes when gravity is read as imbalance.

The cosmos becomes less mechanical and more integrated. Matter, mass, vacuum, gravity, motion, and structure are no longer separate actors forced into relationship. They are different expressions of one deeper field process. The universe is not a warehouse of objects being pulled by mysterious forces. It is a structured field resolving differences through consequences.

This prepares the next step.

If gravity is misunderstood, then the objects created by our misunderstanding of gravity must also be re-examined. Among them, none is more dramatic than the black hole. If gravity is imagined as a force that can collapse matter into infinite density, then the black hole becomes a singularity, a cosmic monster, a place where laws break. But if gravity is energy-gradient behavior within Structured Vacuum Energy, then perhaps the so-called black hole is not a singular thing at all.

Perhaps it is a saturation zone.

Perhaps it is where field coherence fails.

Perhaps it is not the end of physics, but the limit of our current interpretation.

That is where we must go next.

For now, this chapter leaves us with one simple correction.

Gravity does not pull.

Gravity reveals imbalance.

And where imbalance exists, the universe begins its quiet work of returning structure toward balance.

Chapter 6 — Black Holes Don't Exist as We Imagine Them

Why singularities, infinite density, and cosmic monsters may be interpretive errors, and how UED reframes them as Ripple Saturation Zones.

The Monster We Built from Mathematics

Black holes are among the most dramatic ideas ever created by science. They entered the human imagination not merely as astronomical objects, but as cosmic monsters. A region from which nothing escapes. A dark mouth in the universe. A bottomless pit where matter disappears. A place where light fails, time slows, space bends beyond recognition, and physics itself seems to collapse into silence.

The image is powerful.

It is also dangerous.

Because once an idea becomes dramatic enough, we stop questioning the interpretation behind it. We begin to treat the picture as reality. We forget that what we call a black hole is not directly seen in the simple sense of seeing a planet, a star, or a galaxy. We infer it from surrounding behavior, radiation patterns, gravitational effects, motion of stars, energetic emissions, and mathematical models.

Something is there.

Something real is happening.

But what exactly is it?

That question is more important than the name.

Mainstream cosmology and astrophysics have described black holes as regions formed by extreme gravitational collapse. Matter collapses into unimaginable density. An event horizon forms. Beyond that boundary, nothing can return. At the center lies a singularity, a point of infinite density where our known laws of physics break down.

This is the standard story.

But we must pause.

Whenever a theory leads us to infinity, we should not immediately bow down before infinity as if it were nature's final truth. Infinity may not be a discovery. It may be a warning. It may be the point where the model has been stretched beyond its meaningful range. It may be the mathematics saying: beyond this boundary, your current interpretation has lost physical clarity.

A singularity is not an explanation.

It is the failure of explanation wearing a technical name.

Infinite density is not a physical object we can meaningfully imagine. Zero volume containing collapsed mass is not a final understanding. A place where laws break down should not be treated casually as a place in the universe. It may be better understood as a place where our theory has reached its own limit.

This is where Universal Energy Dynamics enters.

UED does not deny that black-hole-like phenomena exist. It does not deny extreme gravitational behavior. It does not deny regions from which ordinary signal recovery becomes impossible. It does not deny the intense centers of galaxies, energetic jets, accretion-like behavior, or the motion of stars around invisible compact regions.

The phenomenon is real.

The interpretation may be wrong.

This distinction is essential.

When this book says, "Black holes don't exist as we imagine them," it does not mean nothing exists where black holes are inferred. It means the cosmic monster may be our construction. The infinite-density singularity may be our mathematical overreach. The bottomless pit may be a metaphor mistaken for mechanism. The black hole may not be a hole at all.

It may be a field condition.

In UED, matter is not an independent substance collapsing into some impossible final point. Matter is ripple condensation within Structured

Vacuum Energy. Mass is a field-density signature. Gravity is not a mysterious force dragging matter into infinite collapse. Gravity is energy-gradient behavior seeking equilibrium.

Once these foundations change, the meaning of black holes must also change.

Extreme gravitational behavior no longer automatically means infinite density. It may mean extreme field saturation. It may mean that ripple coherence has collapsed beyond the ordinary limit of observational recovery. It may mean that the field has reached a boundary where energy condensation, resistance, alignment, and gradient intensity create a zone of darkness—not because reality has ended, but because recoverable structure has failed.

This is why UED reframes black holes as Ripple Saturation Zones.

A Ripple Saturation Zone is not a monster.

It is not a cosmic mouth.

It is not a hole punched in reality.

It is a region where the field becomes so saturated that ordinary ripple information no longer returns in a coherent form to the external observer. What appears as darkness is not necessarily emptiness. What appears as disappearance is not necessarily destruction. What appears as a singularity may be the loss of interpretive coherence.

The monster may not be in the universe.

The monster may be in the model.

Human beings have always created monsters at the edge of understanding. Before science, darkness was filled with demons, gods, spirits, and omens. Modern science replaced those images with equations, but sometimes equations also create their own monsters. When mathematics reaches an extreme and imagination follows without caution, the result can become intellectually beautiful but physically misleading.

The black hole may be one such case.

It may be the place where mathematics, metaphor, fear, and cosmic mystery merged into one powerful image.

But reality may be quieter.

Perhaps the so-called black hole is not where the universe breaks.

Perhaps it is where field coherence reaches saturation.

Perhaps it is not the death of structure, but the regulatory limit of structure.

Perhaps it is not a singular object, but a boundary condition within Structured Vacuum Energy.

This chapter begins from that possibility.

We are not here to deny the darkness.

We are here to understand what the darkness means.

6.1 — How Black Holes Became Cosmic Monsters

Black holes did not begin as monsters.

They became monsters slowly.

First, they entered mathematics. Then they entered astrophysics. Then they entered imagination. And once they entered imagination, they became almost impossible to discuss calmly.

A black hole is now one of the most powerful symbols in modern science. It is not merely an object in space. It is a cultural image. It represents darkness, finality, collapse, fear, mystery, cosmic violence, and the limit of knowledge. Even a child who does not understand general relativity may know that a black hole is something from which nothing escapes.

That sentence itself has enormous power.

Nothing escapes.

Not matter.

Not light.

Not time.

Not information.

Not even our imagination.

This is why black holes became cosmic monsters.

The scientific story begins with gravity. If enough mass collapses into a sufficiently small region, gravity becomes so intense that escape becomes impossible. The idea developed through mathematical extensions of gravitational theory, especially after Einstein's general relativity gave gravity a geometric language. If matter can curve spacetime, then extreme matter concentration can create extreme curvature. Push that curvature far enough, and one reaches a boundary beyond which ordinary return becomes impossible.

This boundary became known as the event horizon.

Beyond it, according to the standard interpretation, the object is hidden from outside observers. The center is imagined as a singularity: a point where density becomes infinite, volume approaches zero, and the known laws of physics fail.

This is already dramatic enough.

But then popular imagination took over.

The black hole became a cosmic drain. A bottomless pit. A dark mouth. A star-eating monster. A place where everything falls and nothing returns. It became the perfect image of the universe's terrifying power. Science gave the equations, but imagination gave the monster its face.

Again, the issue is not that nothing real exists.

Something real is being inferred.

We observe stars moving around invisible compact regions. We observe energetic emissions from matter near such regions. We observe gravitational effects too strong to be explained by ordinary visible objects. We observe galactic centers behaving as if immense compact regulatory zones exist there. These observations cannot be dismissed casually.

The problem is not the phenomenon.

The problem is the interpretation of the phenomenon.

The black hole became a monster because several layers merged into one image: gravitational collapse, event horizon, singularity, infinite density, trapped light, warped spacetime, accretion disks, jets, and public imagination. Each layer added drama. Each layer strengthened the sense that the black hole is not merely a region of extreme field behavior, but a thing — a dark object, a cosmic predator, an ultimate endpoint.

But what if that "thing" is not a thing?

What if the black hole is not an object in the ordinary sense?

What if it is a boundary condition?

What if the so-called monster is a field state where ordinary coherence cannot return to the observer?

This is where the standard image must be examined carefully.

In public language, the black hole is described as if it eats matter. But "eating" is biological language. It belongs to animals, not cosmic field behavior. The black hole is described as if it sucks everything around it. But this too can mislead. Many objects orbit black-hole-like regions without immediately falling in. The dramatic image of a vacuum cleaner in space is not accurate. The region does not chase matter. Matter follows field-gradient conditions.

The monster was created partly because we translated gravitational behavior into emotional language.

Extreme gravity became hunger.

Signal loss became death.

Darkness became emptiness.

Boundary became wall.

Singularity became center.

And mathematical breakdown became cosmic reality.

This is the danger.

When metaphor becomes too strong, it begins to replace mechanism.

A black hole, as popularly imagined, is no longer just an inferred astrophysical region. It becomes a narrative object. It has character. It dominates. It destroys. It swallows. It hides secrets. It becomes the villain of the universe.

But the universe may not contain villains.

It contains conditions.

Universal Energy Dynamics asks us to remove the monster costume and look at the condition beneath. If matter is ripple condensation within Structured Vacuum Energy, then extreme collapse cannot be treated simply as matter falling into an infinitely dense point. If gravity is energy-gradient equilibrium, then extreme gravitational behavior cannot be understood only as attraction becoming infinitely strong. If mass is a field-density signature, then what we call a black hole may represent an extreme field-density condition, not a physical object with impossible internal identity.

This does not make the region less powerful.

It makes it less mythological.

A Ripple Saturation Zone, as UED proposes, is not weak. It is not ordinary. It is not casual. It is an extreme condition where field coherence crosses a threshold. It may prevent recoverable signal return. It may distort surrounding motion. It may regulate galactic structure. It may produce observational darkness. But none of this requires us to imagine a cosmic monster sitting at the center.

The darkness may be real.

The monster may be our interpretation.

This distinction matters because science must not become frightened of its own metaphors. A good metaphor should open understanding. It should not trap intelligence. The fabric metaphor helped explain gravity, but became limiting when treated too literally. In the same way, the black-hole monster helped capture public imagination, but it may now limit deeper interpretation.

The more dramatic the image, the more carefully it must be questioned.

A black hole became a cosmic monster because mathematics reached an extreme, language became dramatic, and imagination filled the darkness with finality. But perhaps the universe is not telling us that reality ends there. Perhaps it is telling us that our current way of reading reality loses coherence there.

The monster may be born not where matter falls.

It may be born where interpretation collapses.

6.2 — The Problem with Singularity and Infinite Density

The word "singularity" sounds scientific.

It sounds precise. It sounds deep. It sounds like a point where science has reached the final secret of reality. But if we listen carefully, singularity is not really an explanation. It is a confession.

It says: our theory has reached a point where its own language no longer works.

In the standard black hole model, matter collapses under extreme gravity. The collapse continues until the matter is compressed into a region of zero volume and infinite density. At that point, the known laws of physics break down. Space and time lose their ordinary meaning.

Curvature becomes infinite. Predictability disappears. The model reaches a mathematical extreme.

This is called singularity.

But what does that really mean?

Can infinite density exist as a physical thing?

Can zero volume hold real matter?

Can a point with no size contain collapsed stars, energy, structure, and history?

Can a place where laws break down be treated as a real location in the universe?

These questions are not emotional objections. They are scientific and philosophical necessities. If a model produces an infinity, we must ask whether nature is truly infinite there or whether the model has been pushed beyond its legitimate domain.

Infinity is often a warning sign.

When calculation gives infinity in physics, it may not always mean the universe contains an infinite object. It may mean our framework has lost resolution. It may mean we have carried a useful equation into a condition where its assumptions no longer apply. It may mean the mathematics has reached a boundary, not the universe itself.

This is why singularity must be handled with great caution.

A singularity is not something we directly observe. We infer black-hole-like regions from surrounding effects: stellar motion, gravitational influence, radiation from nearby material, energetic jets, and boundary-like behavior where ordinary signal recovery becomes impossible. But the singularity itself is not seen. It is the mathematical endpoint of a model.

The danger begins when we convert that endpoint into a physical object.

The model says the collapse continues.

The mathematics says density tends toward infinity.

The imagination says there is a point at the center.

Then the mind begins to treat this point as real.

But perhaps the point is not real.

Perhaps the point is where the model runs out of physical meaning.

This is not a small criticism. It touches the entire mythology of black holes. The singularity gives the black hole its ultimate drama. Without singularity, the black hole becomes less like a cosmic monster and more like an extreme boundary condition. Without infinite density, the language of final collapse loses its power. Without the idea that physics breaks completely at the center, the black hole must be reinterpreted as a state of the field, not a place where reality commits suicide.

Universal Energy Dynamics does not accept infinite density as a final physical explanation.

From the UED perspective, matter is not independent substance that can be crushed endlessly into a mathematical point. Matter is ripple condensation within Structured Vacuum Energy. It is a stabilized field consequence. Mass is not a fixed object-property sitting inside matter. It is a field-density signature. Gravity is not an infinitely hungry force pulling matter into a final pit. It is energy-gradient behavior seeking equilibrium.

Once these foundations change, infinite collapse becomes questionable.

If matter is a field consequence, then extreme compression does not mean matter travels toward infinite identity. It means the field condition approaches saturation. Ripple coherence begins to fail. Alignment collapses. Signal recovery becomes impossible. The observable structure moves toward a limit where the external observer cannot reconstruct what is happening in ordinary terms.

This is not singularity.

This is saturation.

A Ripple Saturation Zone is therefore a more disciplined UED interpretation than an infinite-density point. It does not require us to imagine zero volume. It does not require us to accept physical infinity. It does not require us to treat the breakdown of equations as a real object. It says that the field has crossed a coherence threshold. Beyond this threshold, ordinary ripple information cannot return in a usable form.

The darkness is not necessarily a hole.

It is loss of recoverable coherence.

The event is not necessarily infinite collapse.

It is field saturation.

The mystery is not at the center of a cosmic monster.

It is at the boundary of interpretation.

This shift is important because it protects science from worshipping its own failure points. When a theory says, "Here the laws break down," the honest response should not be, "Then the universe must contain a law-breaking object." The honest response should be, "Our theory has reached its limit; we need a deeper framework."

This is exactly how science matures.

Old models work within certain ranges. Then extreme conditions expose their limits. A new framework does not always erase the old one, but it explains why the old one failed at the boundary. Singularity may be such a boundary.

Newton's gravity worked beautifully until certain observations required deeper interpretation. Einstein expanded the frame. But now black hole singularities may be telling us that even the geometric frame has reached its own edge. If the theory ends in infinity and law-collapse, perhaps the universe is not broken. Perhaps the theory needs a deeper field foundation.

UED offers that field foundation.

It says that extreme gravitational regions must be understood through Structured Vacuum Energy, ripple behavior, field density, alignment threshold, and saturation. The so-called singularity is not the final truth of the black hole. It is the point where a matter-first, spacetime-curvature-based model loses coherence.

In this reading, physical reality does not permit meaningless infinity. The universe is not built on absurdity. It operates through thresholds, limits, balance, saturation, and transformation. When a field condition becomes extreme, the result is not infinite density. The result is a change of state, a collapse of coherence, a boundary of observable recovery.

This is more physically reasonable.

Nature does not need infinity to be powerful.

A finite saturation boundary can be more meaningful than an infinite mathematical ghost. A region where field coherence fails can explain darkness without imagining a point of impossible density. A regulatory zone can explain central galactic behavior without turning it into a cosmic monster.

The singularity, then, may not be a place.

It may be a warning label attached to an incomplete theory.

It may be the model saying, "I cannot describe what happens beyond this limit."

It may be mathematics pointing to the need for a deeper grammar.

This is why UED rejects the worship of infinite density. Not because it rejects science, but because it refuses to confuse mathematical breakdown with physical truth. The universe does not become more scientific when we accept impossibility as an object. Science becomes stronger when it recognizes where interpretation has failed.

Black holes, as observed phenomena, remain important.

But singularities, as infinite-density centers, may belong more to the weakness of the model than to the structure of reality.

Perhaps there is no infinite point.

Perhaps there is no final pit.

Perhaps there is no place where the universe loses its own laws.

Perhaps there is only a field region where ripple behavior reaches saturation, coherence becomes unrecoverable, and external observation names that darkness a black hole.

The singularity may not be the heart of the black hole.

It may be the collapse of our interpretation.

6.3 — Gravity Reconsidered, Black Holes Reconsidered

Once gravity changes meaning, black holes cannot remain the same.

This is unavoidable.

The standard image of a black hole depends heavily on the standard interpretation of gravity. If gravity is imagined as an attractive force that becomes stronger and stronger as mass collapses into a smaller and smaller region, then the idea of an ultimate collapse begins to feel natural. If gravity is imagined as spacetime curvature becoming more and more extreme around compressed mass, then the idea of a deep gravitational pit also feels natural.

In both cases, black holes become the final triumph of gravity.

Matter collapses.

Density increases.

Escape becomes impossible.

Light fails.

A horizon forms.

At the center, the singularity waits.

This story is powerful because it follows the old logic consistently. If mass causes gravity, and gravity grows stronger as mass becomes more concentrated, then enough collapse should produce an extreme region where nothing can return. Push the logic far enough, and the black hole appears.

But what if gravity itself has been misunderstood?

Then the black hole must be reconsidered from the beginning.

In the previous chapter, gravity was not treated as an invisible pull or merely as the bending of empty spacetime. It was reinterpreted as energy-gradient equilibrium within Structured Vacuum Energy. Gravity is the observable behavior of imbalance. Where energy condenses, the field condition changes. Where the field condition changes, gradients arise. Where gradients arise, motion follows the tendency toward balance.

This means gravity is not a separate monster-force waiting to crush matter into infinity.

It is field behavior.

It is structured energy seeking equilibrium.

It is the response of the field to differences in energy density, coherence, resistance, and saturation.

Once this is accepted, extreme gravity no longer automatically means infinite collapse. Extreme gravity means extreme field-gradient condition. It means the field has entered a region of intense imbalance. It means energy condensation has reached a threshold where ordinary coherence may no longer be maintained.

This is a very different reading.

The standard model sees extreme gravitational collapse and asks, "How far inward can matter fall?"

UED asks, "At what point does field coherence saturate?"

The first question leads toward singularity.

The second leads toward Ripple Saturation Zone.

This shift is not a decorative change of terminology. It changes the physical imagination completely. In a matter-first model, the black hole is imagined as collapsed matter concentrated into an impossible center. In UED, the so-called black hole is not primarily a collapsed object. It is an extreme field condition where ripple behavior, energy density, and gradient intensity cross the boundary of recoverable coherence.

In other words, the darkness is not necessarily produced by a thing at the center.

It may be produced by the state of the field.

This also changes the meaning of "collapse." In the standard story, collapse is material. Matter falls inward under gravity. It becomes compressed. It becomes denser. It continues until the model reaches singularity. But under UED, collapse is not merely material compression. It is the collapse of field coherence. The important issue is not how small matter becomes, but how the ripple structure behaves when field intensity becomes extreme.

If ripple behavior remains coherent, signals can resolve.

If coherence weakens, information becomes distorted.

If saturation is reached, recoverable signal structure fails.

From outside, this failure appears as darkness.

That darkness is then interpreted as a black hole.

But the UED reading asks us to be more careful. Darkness does not automatically mean a hole. No-return behavior does not automatically mean infinite density. Signal loss does not automatically mean a singularity. These may all be signs that the field has reached saturation, not that reality has collapsed into an impossible point.

This is why gravity must be reconsidered before black holes can be understood differently.

If gravity remains a pulling force, black holes remain monsters of collapse.

If gravity becomes field-gradient equilibrium, black holes become regions of field saturation.

The difference is enormous.

A monster devours.

A saturation zone regulates.

A monster destroys.

A saturation zone marks a threshold.

A monster is imagined as an object.

A saturation zone is understood as a condition.

This is also why the phrase "black holes don't exist" must be handled carefully. The observed phenomena exist. The gravitational effects exist. The darkness exists. The boundary-like behavior exists. The surrounding energetic activity exists. But the thing imagined beneath these observations may not exist in the way we think.

There may be no infinite-density object.

There may be no bottomless hole.

There may be no singular center where reality ends.

There may instead be a region where Structured Vacuum Energy reaches such intense saturation that ordinary ripple coherence collapses from the observer's frame.

This reading is more consistent with UED because UED does not allow matter, mass, gravity, and vacuum to remain separate categories. Matter is ripple condensation. Mass is a field-density signature. Gravity is gradient behavior. Therefore, an extreme gravitational region must be understood as an extreme field-gradient condition, not simply as matter crushed into absurdity.

In this view, the black-hole-like region is not outside the logic of the universe.

It is a limit expression of the same logic.

It is what happens when energy condensation and gradient intensity reach a threshold where balance cannot be restored through ordinary coherent structure. Instead of visible form, the field produces saturation. Instead of recoverable signal, it produces blackout. Instead of stable matter identity, it produces coherence collapse.

This is why UED can speak of Ripple Saturation Zones.

A ripple is meaningful only when it can carry pattern. When the medium becomes too disturbed, too dense, too resistant, or too saturated, ripple pattern may no longer remain readable. The field may still exist.

Energy may still be present. But the signal does not return in a coherent form. For the outside observer, the region becomes dark, hidden, and extreme.

Mainstream interpretation calls this a black hole.

UED calls it saturation.

This interpretation also helps remove the unnecessary fear built into black hole language. A Ripple Saturation Zone is not a cosmic villain. It is not a supernatural destroyer. It is not a place where the universe commits mathematical violence against itself. It is a field boundary. It is an extreme state of energy-gradient imbalance. It is part of the universe's larger equilibrium system.

This prepares us for an even deeper idea: such regions may not be merely endpoints of collapse. They may function as regulators within galactic systems. If galaxies are field structures, then their centers are not accidental. The central extreme zone may play a role in balancing, stabilizing, redistributing, and regulating the larger galactic field.

That possibility will become important later.

For now, the key point is simple.

Black holes cannot be understood correctly if gravity is misunderstood.

If gravity is treated as a force that pulls matter into infinite collapse, black holes become singular monsters. But if gravity is understood as the field's response to energy imbalance, then black holes become saturation zones where the response has crossed the boundary of ordinary coherence.

The observed darkness remains.

The interpretation changes.

The phenomenon remains.

The monster dissolves.

Gravity reconsidered leads to black holes reconsidered.

6.4 — Ripple Saturation Zones: The UED Reinterpretation

A black hole may not be a hole.

This is the first correction.

The word "hole" itself creates a misleading image. A hole suggests absence. A gap. A tunnel. A pit. A place where something has been removed. When we hear "black hole," the mind imagines a dark opening in space, a cosmic drain where matter falls forever and disappears from reality.

But what if the darkness is not an opening?

What if it is saturation?

Universal Energy Dynamics reframes the black hole as a **Ripple Saturation Zone**. This means it is not an object of infinite density sitting at the center of space. It is not a physical hole punched into the universe. It is not a monster swallowing reality. It is a region where Structured Vacuum Energy reaches an extreme condition of field saturation, and ordinary ripple coherence collapses beyond recoverable observation.

To understand this, we must return to the foundation.

In UED, matter is ripple condensation within Structured Vacuum Energy. Gravity is energy-gradient behavior seeking balance. Mass is a field-density signature. Structure exists when ripple behavior remains coherent enough to stabilize into observable form.

But every coherence has a limit.

A ripple can carry information only when the medium allows the ripple to remain structured. If the medium becomes too disturbed, too dense, too resistant, or too saturated, the ripple no longer returns as readable pattern. The energy may still exist. The field may still exist. The region may still influence surrounding structures. But the signal no longer emerges in a form the external observer can decode.

That is ripple saturation.

In ordinary language, we call such a region dark.

In mainstream cosmology, we call it a black hole.

In UED, we call it a Ripple Saturation Zone.

This term is important because it shifts the discussion away from impossible matter-collapse and toward field-condition limits. The main question is no longer: how can matter collapse into infinite density? The deeper question becomes: what happens when field coherence reaches saturation and recoverable ripple behavior fails?

This is a much cleaner question.

It does not force us to accept infinity as a physical object. It does not require zero volume. It does not require the universe to contain a law-breaking center. It simply says that there are regions where energy-gradient intensity crosses the boundary of ordinary coherence.

Beyond that boundary, observation fails.

Not because reality ends.

But because ripple recovery fails.

This distinction matters deeply.

A dark region is not automatically empty. A silent radio does not mean there is no broadcast anywhere; it may mean the signal is distorted, blocked, saturated, or unrecoverable. A blurred image does not mean no object exists; it means the resolving condition is insufficient. A sound lost in overwhelming noise does not mean silence exists at the source; it means the pattern cannot be separated from saturation.

Similarly, a black-hole-like region may not mean that matter has vanished into a supernatural pit.

It may mean that field behavior has become unrecoverable to the observer.

The darkness is a limit of coherence.

Not necessarily a hole in existence.

A Ripple Saturation Zone can therefore explain why such regions appear extreme without making them absurd. They can have strong external influence because the field-gradient condition around them is intense. They can prevent signal return because ripple coherence fails within the saturated region. They can appear compact because the field condition is highly concentrated. They can regulate surrounding structure because their gradient presence shapes motion and energy redistribution.

None of this requires a singularity.

This is where UED becomes more physically disciplined than the popular black hole imagination. It does not deny the phenomenon; it removes the unnecessary monster. It says: yes, there is an extreme region. Yes, light or information may not return in ordinary form. Yes, surrounding motion may indicate powerful field influence. Yes, galactic centers may contain such zones.

But no, we do not need to imagine infinite density.

No, we do not need to imagine a bottomless hole.

No, we do not need to imagine a place where the universe abandons its own logic.

The universe works through limits.

Chavan's Limit becomes important here. Every manifestation exists between thresholds. Below the lower threshold, the field remains unmanifest or unobservable as stable form. Within the active range, ripple condensation can produce matter, mass, structure, and motion. But when energy density, resistance, or field tension crosses the upper boundary of coherence, ordinary structure can no longer be sustained. The region approaches dissolution or saturation.

A Ripple Saturation Zone is such an upper-threshold condition.

It is not the death of the universe.

It is a boundary of manifest coherence.

This interpretation also changes the meaning of "collapse." Collapse is not merely matter falling inward. Collapse is coherence failing under extreme field conditions. The external observer sees darkness because the ripple structure no longer resolves back outward. The region is not absent; it is saturated. It is not empty; it is over-intense. It is not a hole; it is a field state beyond ordinary recoverable pattern.

This is why the black color matters.

Blackness is not always emptiness. Sometimes blackness is overload. When a detector is saturated, it may fail to produce useful detail. When the eye is overwhelmed or deprived of recoverable light, it sees darkness. Darkness can mean absence, but it can also mean loss of readable interaction.

In UED, the so-called black hole is darkness through saturation.

It is not a vacuum of reality.

It is an excess of unresolved field condition.

This helps explain why these regions can be both invisible and powerful. If they were merely empty holes, their influence would be difficult to understand. But if they are saturated field zones, their surrounding gradient behavior can be enormous. Their darkness and influence are not contradictions. They are two sides of the same condition: internally unrecoverable coherence, externally powerful gradient presence.

This also prepares a more meaningful role for such regions in galaxies.

If a galaxy is a field structure, then its center cannot be treated as an accidental monster. A central Ripple Saturation Zone may function as a regulatory core. It may help balance energy distribution, stabilize rotational behavior, and organize large-scale field coherence. What mainstream language calls a supermassive black hole may actually be the central saturation regulator of a galactic field system.

This idea becomes essential later.

For now, we must understand the reinterpretation clearly.

A black hole is not denied.

The name is questioned.

The observation is accepted.

The interpretation is rebuilt.

Something exists there, but not necessarily as the cosmic monster imagined by inherited language. It may be a Ripple Saturation Zone: a region where field alignment collapses, ripple coherence becomes unrecoverable, and external observation encounters darkness.

This is not less scientific.

It may be more scientific because it avoids worshipping infinity.

It replaces impossible singularity with field threshold.

It replaces cosmic monster with saturation condition.

It replaces hole with limit.

It replaces mystery-as-collapse with mystery-as-field-state.

The black hole, then, may not be where matter disappears from existence.

It may be where structured energy crosses the limit of visible coherence.

And once we understand that, the darkness becomes less like an ending and more like a message:

Here, the field is saturated.

Here, ordinary ripple recovery fails.

Here, interpretation must stop calling it a hole and begin reading it as a threshold.

6.5 — Event Horizon as Observational Boundary, Not Cosmic Wall

The event horizon is one of the most misunderstood ideas in modern cosmology.

In popular imagination, it is often described as a cosmic wall. A boundary around a black hole. A point of no return. Once something crosses it, nothing can come back. Not matter. Not light. Not information. The phrase itself carries a kind of finality. It feels like the universe has drawn a dark circle and declared: beyond this line, reality is hidden forever.

This image is powerful.

But again, power is not the same as precision.

The event horizon may not be a physical wall in space. It may not be a surface like the wall of a room or the edge of a container. It may not be a hard boundary around a cosmic object. It may be better understood as an observational boundary, a limit beyond which ordinary signal recovery becomes impossible for the external observer.

This difference matters.

A wall is a thing.

A boundary can be a condition.

In the standard interpretation, the event horizon is defined by escape. Once inside, the escape velocity exceeds the speed of light. Since nothing can travel faster than light, no signal can return. This creates the blackness. This creates the mystery. This creates the idea that whatever falls beyond the horizon is cut off from the outside universe.

Universal Energy Dynamics does not deny the boundary-like behavior.

It questions its meaning.

From the UED perspective, the event horizon is not necessarily the edge of a hole. It is the outer limit of recoverable ripple coherence. It marks the region where field saturation becomes so intense that ordinary signal structure cannot resolve back to the external observer. Light may not return as readable light. Information may not return as coherent information. Ripple behavior may be absorbed, distorted, saturated, or dissolved beyond recognition.

So the event horizon is not a magical wall.

It is a coherence boundary.

This is a more careful interpretation.

If matter is ripple condensation within Structured Vacuum Energy, then every observable signal is a form of field behavior. Light, radiation, motion, and gravitational effects are not independent messages flying through empty nothingness. They are consequences of energy interacting with the structured field. For a signal to be observed, it must retain enough coherence to be resolved.

But inside a Ripple Saturation Zone, coherence fails.

The field condition becomes too intense. Energy density, resistance, and gradient saturation cross the threshold where ordinary ripple patterns can no longer maintain recoverable structure. To the outside observer, this appears as a limit. Not because there is necessarily a physical wall, but because the field no longer returns readable pattern from beyond that boundary.

This is why the event horizon may be better understood as the horizon of interpretation.

It is where observation loses access.

It is where the signal cannot translate itself back into the language of external measurement.

It is where the universe does not disappear, but becomes unreadable through ordinary channels.

This is similar to standing at the edge of dense fog. The fog does not contain a wall. There is no solid surface where visibility ends. Yet beyond a certain distance, the eye cannot recover detail. The world still exists beyond the fog, but visibility fails. The boundary belongs to the condition of observation, not to a physical barrier in the world.

The analogy is limited, but useful.

The event horizon may be the cosmic version of such a visibility boundary. Not fog in space, not physical mist, but field saturation. A region where ripple information cannot return in a coherent enough form to be recognized.

This also helps us avoid unnecessary mythological language.

When we say something "falls into a black hole," the mind imagines an object crossing a rim and disappearing into a pit. But under UED, what may be happening is not falling into a hole, but entering a region of extreme field-gradient saturation. The object's matter-structure, already understood as ripple condensation, encounters a field condition where its coherence cannot remain externally recoverable. Its identity, signal, and structure no longer translate back to outside observation.

The disappearance is observational.

The transformation is field-based.

This is why the event horizon should not be treated as proof of singularity. A boundary of no return does not automatically prove infinite density at the center. It only tells us that beyond a certain field condition, signals fail to return. What lies deeper may not be a singular point. It may be further saturation, transformation, redistribution, or collapse of coherent ripple identity.

The event horizon marks loss of access.

It does not necessarily reveal the final nature of the interior.

This is a crucial scientific discipline. We must not convert observational limitation into physical certainty. When we cannot see beyond a boundary, we should not immediately fill the unseen region with infinite density. We should admit that the boundary tells us something about signal recovery, not everything about ultimate reality.

In UED, the event horizon is therefore a threshold.

It is a Chavanian boundary of recoverable coherence.

Before the boundary, ripple behavior can still produce observable signals. Around the boundary, signals become distorted, delayed, stretched, weakened, or trapped in extreme gradient conditions. Beyond the boundary, coherent recovery fails. The external observer receives darkness not because reality is absent, but because the field has crossed into saturation.

This understanding also changes the emotional meaning of "point of no return."

No return does not mean cosmic death in the dramatic sense. It means no return of ordinary coherence to the observer's frame. The field may still process energy. It may still redistribute imbalance. It may still maintain regulatory influence. But the external observer cannot receive normal signal structure from within that condition.

This is why black-hole-like regions can be both invisible and influential.

Their internal ripple coherence is unrecoverable, yet their surrounding gradient remains powerful. They do not communicate ordinary light from inside, but they shape motion, structure, and energetic behavior around them. This dual nature is not mysterious under UED. It is exactly what a saturation boundary would produce.

Inside: unrecoverable coherence.

Outside: strong field-gradient influence.

The event horizon becomes the transition between these two regimes.

It is not a wall.

It is a boundary between readable and unreadable field behavior.

This also allows us to reinterpret the phrase "nothing escapes." In UED language, the better statement may be: ordinary coherent ripple information does not recover beyond the saturation boundary. That is less dramatic, but more precise. It removes the monster language and replaces it with field language.

Nothing escapes sounds like imprisonment.

Coherence fails sounds like physics.

This distinction is important for the whole book. The aim is not to make the universe less powerful. It is to make our interpretation less theatrical. The event horizon does not become less significant when reinterpreted. It becomes more meaningful. It is no longer a dark wall around an impossible object. It is the measurable edge of a field condition where observation loses coherence.

This is why the event horizon may be one of the most important clues, not to the existence of a singularity, but to the limits of field resolution.

A horizon is not the end of the world.

When we stand on Earth and see the horizon, we do not think the world ends there. We understand that our view ends there. The horizon is real as an observational limit, not as the edge of existence.

Perhaps the event horizon should be read in the same spirit.

It is real as a boundary of recoverable signal.

But it may not be the edge of reality.

It may be the place where our ordinary observational relationship with the field ends.

The region beyond may not contain a hole, a monster, or an infinite-density point. It may contain a field state that our current language cannot describe properly. UED gives us a better beginning: Ripple Saturation Zone. Within that zone, coherence is no longer available to external observation in the usual way.

This does not solve every detail. It opens a more disciplined direction.

Instead of asking what happens inside a magical pit, we ask what happens to ripple coherence under extreme saturation. Instead of asking how matter reaches infinite density, we ask how field identity transforms beyond the recoverable threshold. Instead of treating the event horizon as a wall, we study it as a transition in field behavior.

That is a more useful path.

The event horizon may not be the boundary of the black hole as an object.

It may be the boundary of our ability to read the field.

And if that is true, then the darkness beyond it does not say, "Reality ends here."

It says, "Coherence no longer returns in your language."

6.6 — Central Regulators, Not Cosmic Endpoints

A black hole is usually imagined as an endpoint.

A star collapses. Matter falls. Light disappears. Gravity wins. The story ends in darkness. In this imagination, the black hole is the final destination of matter, the last stage of collapse, the place where structure goes to die.

But what if this is the wrong emotional frame?

What if the so-called black hole is not the death of structure, but the regulator of structure?

This is one of the most important shifts in the UED reinterpretation.

If a galaxy is understood as a field structure, then its center cannot be treated as an accidental monster. The center is not merely a graveyard where matter disappears. It may be the regulatory core of the entire galactic field configuration. It may be the zone where extreme field gradients, ripple saturation, energy redistribution, and structural coherence meet.

In mainstream cosmology, supermassive black holes are often treated as objects sitting at the centers of galaxies. They are massive, compact, gravitationally dominant, and influential. Their relationship with galaxies creates difficult questions. Did the galaxy form first and then produce the black hole? Did the black hole form first and help build the galaxy? How did such massive central objects appear so early in the universe? How did they grow fast enough?

This becomes the famous chicken-and-egg problem.

Which came first: the galaxy or the black hole?

But perhaps this question is created by the wrong framework.

It assumes that the galaxy and the central black hole are two separate objects formed in sequence. First one appears, then the other develops. One must precede the other. One must cause the other. The problem becomes chronological because the model begins with separate things.

Universal Energy Dynamics begins differently.

In UED, a galaxy is not merely a collection of matter. It is a field consequence. It is structured energy expressed through ripple condensation, gradient behavior, and equilibrium-seeking organization. If this is true, then the central extreme region is not a separate object added later to the galaxy. It may be part of the same field-realignment event.

The galaxy and its central regulator may arise together.

This removes the chicken-and-egg problem.

A galactic field system may require a central regulatory condition from the beginning of its manifestation. Just as a cyclone has an eye, a whirlpool has a central structure, and a rotating system has an organizing axis, a galaxy may require a central saturation-regulation zone to maintain large-scale coherence. The center is not an accident. It is not merely a monster sitting inside the structure. It is part of how the structure remains structured.

This does not mean the center behaves gently.

A central regulator can be extreme. It can be intense. It can dominate the surrounding field. It can influence motion, radiation, redistribution, and stability. It can create powerful jets, energetic emissions, and strong gravitational effects. But intensity does not automatically mean destruction. In many systems, the most intense region is also the region of regulation.

The heart is intense, but it sustains circulation.

The nucleus is small, but it organizes the atom.

The eye of a storm is central to the storm's structure.

A central regulator may appear dangerous from one angle and necessary from another.

The same may be true of black-hole-like regions in galaxies.

From the outside, they appear dark, extreme, and unreachable. Mainstream language calls them supermassive black holes. UED calls them central Ripple Saturation Zones. They are regions where field coherence collapses beyond ordinary recoverable observation, but their surrounding gradient behavior helps define the larger galactic system.

This is why they should not be treated merely as cosmic endpoints.

They may be cosmic regulators.

A galaxy must manage motion, energy distribution, rotational coherence, star formation patterns, field gradients, and large-scale structural stability. If the galaxy is a field consequence, then this management cannot be explained only through scattered stars and gas clouds. There must be a regulatory geometry inside the field. The central saturation zone may serve this function.

It becomes the place where excess field tension is absorbed, redistributed, or balanced.

It becomes the region where extreme gradients are concentrated.

It becomes the boundary condition that helps organize the larger field.

It becomes the dark core around which visible structure maintains coherence.

Under this interpretation, the darkness is not absence.

It is regulatory saturation.

This idea also helps reinterpret JWST's early supermassive black hole problem. In the standard timeline, supermassive black holes appearing at cosmic depth are difficult because they seem too large too soon. They could not have grown through ordinary stellar collapse and gradual accretion within the allowed time. So the model requires heavy seeds, direct collapse scenarios, unusually rapid growth, or exceptional feeding conditions.

But UED asks another question.

What if they did not grow chronologically as separate objects?

What if central regulators arise with galactic field structures?

If a galaxy manifests as a field-alignment system, then its central saturation zone may not need to wait for billions of years of feeding. It may be established as part of the same structural blueprint. The galaxy is not first built and then given a regulator. The regulator is part of the field condition that allows the galaxy to be coherent in the first place.

This is a major conceptual shift.

A central regulator does not need to be a cosmic infant that somehow became gigantic too fast. It is not necessarily an object that skipped childhood. It may be the central saturation condition of a large-scale field system, appearing with the system because the system requires it.

This makes the so-called early supermassive black hole less anomalous.

The problem was created by imagining it as a separate object that had to grow inside a limited timeline. If instead it is understood as a regulatory field zone born with the galactic configuration, the chronological difficulty weakens.

The central zone is not late.

It is structural.

This also changes how we understand the relationship between galaxies and their centers. In the old view, the galaxy and black hole interact as two entities. The black hole feeds on matter. It affects star formation. It may grow with the galaxy. Their masses may correlate. The relationship is real, but it is described as interaction between separate objects.

In UED, the relationship is more unified.

The galaxy and its central regulator are two expressions of one field architecture. The visible galaxy expresses ripple condensation and gradient organization across large scale. The central regulator expresses saturation and boundary behavior at the core. One radiates structure outward. The other anchors the gradient inward. Together, they form a coherent field system.

This is why the word "regulator" is important.

It does not mean a conscious controller. It does not mean an intelligent machine sitting at the center. It means a structural function within the field. The central zone regulates because its condition influences the balance, motion, and coherence of the surrounding system.

In a living body, regulation does not always mean command. Temperature regulation, blood pressure regulation, hormonal regulation, and neural regulation all involve feedback, balance, thresholds, and response. Similarly, in a galaxy, regulation may involve energy gradients, saturation thresholds, rotational coherence, and redistribution behavior.

A central Ripple Saturation Zone may be the galaxy's field-regulatory threshold.

It may prevent uncontrolled dispersion.

It may absorb excess disturbance.

It may anchor rotational coherence.

It may maintain gradient balance.

It may define the structural identity of the galactic system.

This does not make the galaxy biological. It simply shows that regulation is a universal principle of organized systems. Wherever structure exists, balance must be maintained. Wherever balance is disturbed, regulation appears. In UED, regulation is not an added feature. It is a consequence of structured energy seeking equilibrium.

This is why black-hole-like regions should be reinterpreted.

They are not cosmic endpoints where reality dies. They may be boundary regions where the galaxy maintains its deeper field identity. They are not monsters eating structure. They may be saturation regulators allowing structure to persist. They are not law-breaking centers. They may be threshold conditions that reveal the upper limits of field coherence.

Once this is understood, the emotional image changes completely.

The black hole is no longer the villain at the center of the galaxy.

It becomes the silent regulator.

The darkness is no longer merely frightening.

It becomes functional.

The center is no longer only a place of collapse.

It becomes a place of balance under extreme condition.

This prepares the next chapter naturally. If central Ripple Saturation Zones are regulators, then galaxies and their cores should not be studied as separate chronological products. They may arise together as unified field-realignment events. The galaxy may not create the regulator later, and the regulator may not create the galaxy alone. Both may be born as one structured consequence of the field.

The so-called black hole, then, may not be the end of cosmic structure.

It may be the condition that allows cosmic structure to remain coherent.

The universe is not placing monsters at the centers of galaxies.

It may be placing balance points there.

Closing Reflection — The Darkness May Not Be Empty

Darkness is easy to misunderstand.

When we cannot see something, the mind quickly fills the unknown with fear, mystery, or finality. Darkness becomes absence. Silence becomes emptiness. No return becomes destruction. A boundary becomes a wall. A limit of observation becomes a limit of reality itself.

This is exactly what may have happened with black holes.

We saw darkness.

We inferred extreme gravity.

We built mathematics around collapse.

We imagined a horizon.

We accepted singularity.

Then we allowed the imagination to complete the monster.

But perhaps the darkness was never empty.

Perhaps the darkness was never a hole.

Perhaps the darkness was the field reaching a condition where ordinary coherence no longer returned to us.

This is the central correction of this chapter.

The observed phenomenon is not denied. Something real exists at the centers of galaxies and in regions where extreme gravitational behavior is inferred. Stars move around invisible compact centers. Energetic emissions arise around them. Matter behaves strangely near them. Light fails to return in ordinary ways. The evidence points to extreme conditions.

But extreme condition does not automatically mean infinite density.

It does not automatically mean a singularity.

It does not automatically mean a cosmic monster.

It may mean saturation.

A Ripple Saturation Zone gives us a different language. It allows us to speak of darkness without turning it into a hole. It allows us to speak of no-return behavior without making it a magical prison. It allows us to speak of extreme gravitational effects without worshipping infinity. It allows us to reinterpret the so-called black hole as a field condition where ripple coherence collapses beyond recoverable observation.

This is not a small change.

It changes the emotional and scientific identity of the black hole.

The black hole is no longer the place where the universe breaks. It becomes the place where our ordinary reading of the field fails. The singularity is no longer treated as the heart of reality. It becomes a warning that the model has reached its interpretive limit. The event horizon is no longer a cosmic wall. It becomes the boundary of recoverable coherence. The galactic center is no longer merely a destructive endpoint. It becomes a possible regulatory core within a larger field structure.

In this view, darkness has function.

It is not dead absence.

It is not meaningless collapse.

It is not cosmic punishment.

It may be the signature of a field state so intense that ordinary visibility cannot survive. The field is still present. Energy is still present. Influence is still present. But the form of recoverable information changes. What returns to us is not light, not image, not ordinary structure, but effect.

Motion tells us something is there.

Gradient tells us something is there.

Surrounding behavior tells us something is there.

But the inner coherence does not return.

That is not emptiness.

That is saturation.

This is why UED refuses to treat black holes as final answers. The phrase "black hole" has become too loaded. It carries the weight of metaphor, mathematics, fear, and cultural imagination. It makes us think we already know what the darkness means. But perhaps the name has become heavier than the phenomenon.

A better science must be willing to rename when necessary.

Not for fashion.

Not for rebellion.

But because wrong names create wrong thinking.

If we call it a hole, we imagine absence.

If we call it a monster, we imagine destruction.

If we call it a singularity, we accept mathematical breakdown as physical truth.

But if we call it a Ripple Saturation Zone, we begin asking better questions.

What field condition produces saturation?

Where does ripple coherence fail?

How does energy redistribute near the boundary?

How does the central saturation zone regulate galactic structure?

What does the event horizon really mark?

How does the surrounding field remain influenced when internal coherence becomes unrecoverable?

These questions do not end the mystery.

They discipline it.

They remove theatrical fear and replace it with field-based inquiry. They allow the darkness to become scientifically meaningful again. They open the possibility that what we have called the most destructive object in the universe may actually be one of the most important regulatory conditions in cosmic structure.

That is a profound reversal.

The black hole may not be the end of matter.

It may be the limit of matter's coherent expression.

It may not be the death of light.

It may be the failure of ripple recovery.

It may not be the collapse of law.

It may be the boundary where our current laws lose language.

It may not be where reality ends.

It may be where interpretation loses access.

This prepares us for the next step. If these dark central regions are not monsters but regulators, then galaxies must also be reinterpreted. Their centers cannot be treated as accidental pits. Their structures cannot be treated as separate from their cores. The galaxy and the central regulator may be two expressions of one field architecture.

The visible galaxy may be the outward structure.

The dark center may be the inward boundary.

Together, they may form a unified equilibrium system.

This is why the darkness matters.

It may be silent, but it is not meaningless.

It may be invisible, but it is not absent.

It may be unrecoverable, but it is not unreal.

The black hole, as imagined, may not exist.

But the field condition behind it may be one of the deepest realities in the cosmos.

So we must be careful before calling darkness empty.

Sometimes darkness is not the absence of reality.

Sometimes darkness is reality beyond the coherence of our present interpretation.

Chapter 7 — The Central Regulator of Galaxies

How galaxies and their regulatory cores may arise together as unified field-realignment events.

The Center Was Never an Accident

A galaxy is usually imagined from the outside.

We see its arms, its glow, its stars, its dust lanes, its spiral beauty, its enormous spread across cosmic space. The mind naturally begins with the visible. A galaxy appears as a grand luminous city of stars, rotating in silence, stretched across darkness like a cosmic design painted on the vacuum.

Then, somewhere at the center, we place the black hole.

As if the galaxy is one thing and the center is another.

As if the bright structure came first, and later a dark object arrived to occupy the middle.

As if the center is merely a hidden tenant inside a house already built.

But perhaps this is the wrong way to see a galaxy.

Perhaps the center was never an accident.

Perhaps the center was never merely an object sitting inside the galaxy. Perhaps it is not a leftover of collapse, not a cosmic monster, not a dark intruder placed at the heart of light. Perhaps the center is part of the very condition that allows the galaxy to remain a galaxy.

This is the shift we must make.

In the older imagination, a galaxy is treated as a collection. Stars, gas, dust, matter, dark matter, motion, gravity, and time slowly work together.

Somewhere inside this long mechanical story, a central black hole grows. It consumes matter. It gains mass. It becomes supermassive. Then scientists must ask the difficult question: which came first, the galaxy or the black hole?

But this question may already contain the mistake.

It assumes that the galaxy and the central black-hole-like region are two separate things with two separate histories. It assumes one must be older and the other younger. It assumes one must produce the other. It assumes a chronological sequence where perhaps there is actually a unified field event.

Universal Energy Dynamics asks us to look differently.

If matter is ripple condensation within Structured Vacuum Energy, and gravity is energy-gradient behavior seeking balance, then a galaxy is not merely a pile of stars. It is a field architecture. It is a large-scale pattern of structured energy, expressed through condensation, rotation, luminosity, gradient, and equilibrium.

In such an architecture, the center is not decorative.

The center is regulatory.

A whirlpool is not simply water moving around an empty middle. Its center is part of its identity. A cyclone is not only wind spread across a region; its eye belongs to its structure. An atom is not merely electrons moving randomly; its nucleus defines the field of the atom. A living body is not merely limbs and organs; its regulatory systems hold the organism in functional balance.

A galaxy, too, may require a central regulatory condition.

Not because it is alive like a body.

Not because it has intention like a mind.

But because every stable structure needs a way to manage imbalance.

Where there is rotation, there must be relational balance. Where there is energy distribution, there must be regulation. Where there is field tension, there must be a place or condition through which excess, saturation, and gradient behavior are managed. Without such a regulatory core, large-scale coherence may not sustain itself.

This is why the so-called black hole at the center of a galaxy may need to be reinterpreted.

It may not be the death point of the galaxy.

It may be the balancing condition of the galaxy.

It may be a Central Ripple Saturation Zone, a region where field coherence reaches saturation and ordinary signal recovery fails, but where the surrounding gradient helps organize the larger galactic structure. Its darkness may not mean absence. Its silence may not mean emptiness. Its no-return behavior may not mean destruction. It may mean that the field has crossed a regulatory threshold.

The visible galaxy shines outward.

The dark center holds inward.

Together, they may form one system.

This is why the center cannot be treated as an afterthought. If the galaxy is a unified field-realignment event, then the regulator and the luminous structure arise together. The galaxy does not first become a galaxy and then later search for a center. The center is part of the very field condition through which the galaxy becomes coherent.

This also changes the problem of early supermassive black holes.

If we imagine them as objects that must grow slowly by feeding, then their early appearance becomes difficult. There seems to be too little time. Too much mass. Too much maturity. Too much central power. The old model must then invent special seeds, rapid growth, or unusual feeding conditions.

But if the central regulator arises with the galactic field architecture, the difficulty weakens.

It does not need to grow as a separate monster after the galaxy appears. It may be established as the inward regulatory boundary of the same field event that expresses the galaxy outwardly.

The galaxy and the regulator are not two separate stories.

They may be one structure seen from two sides.

The luminous side shows ripple condensation, star formation, motion, and visible architecture.

The dark side shows saturation, boundary, gradient concentration, and regulatory coherence.

This is the deeper meaning of the central regulator.

A galaxy may not shine despite its darkness.

It may shine because its darkness holds the pattern.

7.1 — The Chicken-and-Egg Problem of Galaxies and Black Holes

One of the most persistent puzzles in modern cosmology is the relationship between galaxies and the massive dark regions at their centers.

The question appears simple.

Which came first?

Did the galaxy form first and later create the supermassive black hole at its center? Or did the black hole form first and help gather matter around itself to build the galaxy? Did stars, gas, dust, and gravitational structure slowly assemble, eventually feeding the central object? Or did the central object act as the original anchor around which the galaxy organized?

This is the chicken-and-egg problem of galaxies and black holes.

At first, it sounds like a technical issue. But beneath it lies a much deeper problem of interpretation. The question exists because the galaxy and its center are treated as two separate things. One is imagined as a luminous structure made of stars, gas, and matter. The other is imagined as a dark compact object sitting inside it. Once they are separated in thought, chronology becomes unavoidable. One must come before the other. One must produce, feed, grow, influence, or regulate the other.

The mind then searches for sequence.

If the galaxy came first, then it must explain how enough matter reached the center to form and grow a supermassive black hole. But this becomes difficult when massive central black-hole-like regions are inferred in very deep, high-redshift systems. There seems to be too little time for ordinary stellar collapse, feeding, merging, and gradual growth.

If the black hole came first, another problem appears. How did such an extreme central object form before the galaxy itself? What created the seed? How did it become massive enough? How did it avoid destroying or disturbing the very structure it supposedly helped form?

Either way, the old framework struggles.

So modern cosmology introduces possible solutions. Perhaps there were heavy black hole seeds. Perhaps some early massive stars collapsed directly. Perhaps gas clouds formed black holes without going through the usual stellar pathway. Perhaps accretion was unusually efficient. Perhaps mergers happened rapidly. Perhaps early cosmic conditions allowed exceptional growth.

These are serious attempts, and they deserve examination.

But they still remain trapped inside the same basic assumption.

They still treat the galaxy and the central black hole as separate entities with separate formation histories.

This is where Universal Energy Dynamics changes the question.

UED does not begin with two objects. It begins with one field condition.

If a galaxy is not merely a collection of stars but a large-scale field architecture, then the central dark region is not an object added later. It is part of the same architecture. The galaxy and its central regulator may not be two independent things competing for chronological priority. They may be two expressions of one field-realignment event.

In this view, the chicken-and-egg problem weakens because the chicken and egg were never fully separate.

The visible galaxy is the outward expression of ripple condensation, rotational coherence, energy distribution, and structured field manifestation. The central regulator is the inward expression of saturation, gradient concentration, coherence boundary, and field regulation. One shines outward. The other holds inward. Both belong to the same system.

The mistake begins when we treat the dark center as a cosmic object sitting inside a luminous object.

A galaxy is not a container.

The center is not cargo.

A galaxy is a field pattern, and its center is the regulatory condition of that pattern.

This becomes easier to understand through simpler systems. A whirlpool does not first form as rotating water and later install a center. The center is part of the whirlpool's structure from the beginning. A cyclone does not first become a storm and later create an eye as an accessory. The eye belongs to the storm's organization. An atom does not first appear as an electron cloud and later decide to include a nucleus. The nucleus is central to the field identity of the atom.

These examples are not perfect copies of a galaxy, but they reveal an important principle: organized systems often do not create their centers later as separate accidents. Their centers are part of their structural identity.

A galaxy may be similar.

Its central Ripple Saturation Zone may arise with the galactic field itself because large-scale coherence requires a regulatory boundary. The visible arms, star-forming regions, rotational motion, and luminous spread may require an inward saturation condition to anchor and balance the system. Without such a regulator, the galaxy may not maintain its structure as a coherent field arrangement.

This is why early supermassive black holes become so important.

In the standard model, they look problematic because they appear too massive too soon. But under UED, they may not be "early monsters" at all. They may be central regulators appearing with their galactic field structures. Their so-called early maturity may not be a sign of impossible growth. It may be a sign that the galaxy and regulator emerged together as a unified field event.

The question therefore changes.

We no longer ask only, "How did the black hole grow so fast?"

We ask, "Why are we assuming it had to grow separately?"

We no longer ask only, "Did the galaxy come first or the black hole?"

We ask, "Are they two separate histories, or one field architecture?"

This is the key shift.

The chicken-and-egg problem may not be a real cosmic paradox. It may be a conceptual artifact created by object-based thinking. We divided one field system into two objects, placed them on a timeline, and then became confused about which object came first.

But the universe may not have divided them that way.

The galaxy and its central regulator may be born together, not as two objects, but as two complementary expressions of one structured energy condition.

The luminous structure reveals the field becoming visible.

The dark center reveals the field reaching regulatory saturation.

Together, they form the galaxy.

7.2 — Why Every Large Structure Needs a Regulatory Core

No large structure remains stable by accident.

Wherever there is organization, there is regulation. Wherever there is motion, there is balance. Wherever there is energy distribution, there is some form of correction, adjustment, anchoring, or threshold control. Without regulation, structure does not remain structure for long. It disperses, collapses, overheats, fragments, or loses coherence.

This is true in ordinary systems.

A living body survives because it regulates temperature, blood pressure, breathing, digestion, hormones, impulses, and countless internal processes. The body is not merely a collection of organs. It is a regulated arrangement. If regulation fails, the organs may still exist, but the body cannot function as a coherent system.

A cyclone also has regulation. It is not merely random wind. It has rotation, pressure difference, flow structure, and a central eye. The eye is not a decorative emptiness. It belongs to the storm's organization. Remove the field condition that sustains the eye, and the storm itself changes.

A whirlpool has a center. The center is not an object placed inside the water. It is the organizing condition of the rotational flow. The visible motion around it makes sense because the entire system has a central relation. Without that central condition, the whirlpool dissolves into ordinary movement.

Even an atom reveals the same principle. The atom is not a loose crowd of particles. It has internal structure, relational balance, and a central nuclear condition around which the whole identity becomes meaningful. Again, this is not because the atom is alive or conscious. It is because structure requires organization, and organization requires a regulatory principle.

A galaxy is far larger, but the principle does not disappear.

A galaxy contains stars, gas, dust, radiation, motion, rotation, gravitational behavior, energetic exchange, and large-scale structure. Such a system cannot remain coherent if everything is only scattered movement.

There must be deeper order. There must be a way for motion, density, field tension, and energy distribution to remain structurally related.

In mainstream cosmology, this regulation is often assigned to gravity, dark matter halos, and central black holes. But Universal Energy Dynamics asks whether these are separate ingredients or surface expressions of one deeper field architecture.

From the UED perspective, a galaxy is not merely a collection needing support. It is a field structure expressing itself through ripple condensation, energy gradients, rotational coherence, and equilibrium-seeking behavior. If this is true, then the galactic center is not accidental. It is not merely a place where matter happened to fall. It may be the regulatory core of the entire structure.

This regulatory core does not need to be imagined as a conscious controller.

Regulation does not always mean command. In nature, regulation often means threshold behavior. It means balance through response. It means a system managing difference, pressure, tension, flow, and instability. The regulator is not necessarily a ruler. It is a condition that keeps the system from losing coherence.

A central Ripple Saturation Zone may serve such a function in a galaxy.

It may concentrate extreme field gradients. It may absorb or contain excess disturbance. It may define the inward limit of coherence. It may act as the saturation boundary around which the larger field structure maintains its rotational and energetic identity. What looks like a dark destructive center may actually be a necessary balancing condition.

This is why calling it only a black hole may be misleading.

The phrase "black hole" makes the center sound like an endpoint. It suggests collapse, disappearance, and destruction. But a regulatory core has a different meaning. It is not merely where things end. It is where the system's imbalance is managed. It is where the field reaches a saturation condition that helps define the structure around it.

The visible galaxy may shine because the invisible center holds the gradient.

Without a regulatory core, a large-scale field structure would have no inward reference. Its motion may lose coherence. Its gradients may fail to stabilize. Its energy distribution may become uncontrolled. Its rotational behavior may not sustain the same organized pattern.

This does not mean every detail is already solved. It means the direction of interpretation changes.

Instead of asking only how a black hole grew at the center of a galaxy, UED asks why a galaxy would exist without a central regulatory condition in the first place. Instead of treating the center as an object that formed later, it treats the center as part of the system's structural necessity.

Large structures need regulation because large structures carry large imbalance.

A galaxy is not a peaceful decoration in space. It is a massive field arrangement. It contains motion, density, luminosity, collision, radiation, formation, dissolution, and redistribution. Such a structure cannot remain stable if the field does not have a way to manage its own internal tensions.

This is where the central regulator becomes essential.

It is the dark balancing point of the luminous system.

It is the inward boundary of outward structure.

It is the saturation condition that prevents the galaxy from being merely a scattered crowd of stars.

In UED language, structure without regulation cannot sustain coherence.

And a galaxy, being one of the largest coherent structures we observe, may require its central regulator not as an accident, but as a condition of its very existence.

7.3 — The Galaxy as a Field Architecture

A galaxy is not a pile.

This may be one of the most important corrections we need to make.

When we see a galaxy, the mind naturally sees a collection: billions of stars, clouds of gas, dust lanes, radiation, motion, and a center. Because the parts are visible or inferable, we assume the galaxy is made by gathering those parts together. It becomes a cosmic crowd, a large population of objects held in place by gravity.

This view is useful at the surface level.

A galaxy does contain stars. It does contain gas. It does contain dust. It does show motion, rotation, luminosity, and large-scale shape. These are real features. But the danger begins when we treat these features as the final explanation of the galaxy. A crowd of stars does not automatically explain why the galaxy behaves as a coherent structure. A collection of parts does not automatically explain why there is rotation, shape, gradient, center, stability, and long-term field identity.

The galaxy is more than its contents.

It is an architecture.

But not an architecture built from bricks.

It is a field architecture.

Under Universal Energy Dynamics, a galaxy is understood as a large-scale expression of Structured Vacuum Energy. Matter within it is ripple condensation. Mass is field-density signature. Gravity is energy-gradient behavior. Rotation is not merely mechanical motion; it is a dynamic expression of field balance. Luminosity is not merely scattered starlight; it is the visible output of condensed energy structures. The center is not merely a dark object; it is the regulatory threshold of the whole arrangement.

This means the galaxy must be read as a unified field pattern.

Its stars are not isolated lamps floating in emptiness. They are local condensations within a wider field condition. Its arms are not accidental trails of matter. They are visible patterns of large-scale field organization. Its rotation is not merely the result of objects moving around a center. It is the behavior of a gradient system maintaining coherence. Its center is not a separate object inserted into the middle. It is the inward regulatory condition of the field architecture.

This changes the question completely.

Instead of asking only, "What is inside the galaxy?" we must ask, "What field condition allows these contents to remain organized as a galaxy?"

This is the deeper question.

A house is not understood merely by counting bricks. A living body is not understood merely by listing organs. A song is not understood merely by naming notes. The arrangement matters. The relational order matters. The hidden coordination matters. Similarly, a galaxy cannot be fully understood by counting stars, estimating gas, adding dark matter, and measuring rotation. These are important, but they do not fully explain the organizing field.

The galaxy has geometry.

It has rhythm.

It has balance.

It has inward and outward relation.

It has visible and invisible coherence.

In UED, this coherence arises from field realignment. A galactic system appears when Structured Vacuum Energy reaches a condition where ripple condensation, energy gradients, rotational behavior, and regulatory saturation organize into a stable large-scale pattern. The galaxy is not first a loose crowd and then later a structure. It is a structure from the moment the field condition expresses itself coherently.

This is why the central regulator matters.

If the galaxy is a field architecture, then the center is part of the architecture. It is not an optional feature. It is not merely a collapsed object sitting at the middle. It is the inward boundary where field intensity reaches saturation. It is the region where ordinary coherence becomes unrecoverable but surrounding gradient behavior remains powerful. It is the dark regulatory core around which the luminous field pattern maintains itself.

The visible galaxy spreads outward.

The central regulator anchors inward.

Both belong to the same architecture.

This removes the need to imagine the galaxy and its center as two separate histories. The galaxy does not first appear as a bright structure and then later develop a dark core like an afterthought. The core is part of the field condition that allows the bright structure to remain coherent. It is like the center of rotation in a whirlpool. The visible motion and the central condition are not separate events. They are one system.

This also helps explain why galactic centers are so powerful.

A regulatory core must carry intensity. It must manage imbalance. It must concentrate gradients. It must define a boundary between recoverable coherence and saturation. Its darkness is not weakness. Its invisibility is not emptiness. Its no-return behavior may be the very sign of its regulatory role.

The mistake is to treat darkness as destruction only.

In a field architecture, darkness may be boundary.

It may be containment.

It may be regulation.

It may be the silent condition that allows visible structure to persist.

This is why galaxies should not be imagined as matter floating inside empty space. The space between stars is not meaningless. The vacuum is not a passive gap. The field is continuous. The visible stars are only local expressions of a deeper structure. What we call the galaxy includes not only what shines, but also the field condition that holds the shining together.

The galaxy is therefore not merely visible matter plus hidden matter.

It is a coherent energy-field system.

This view also changes how we interpret JWST observations. When deep-field galaxies appear mature, compact, bright, or centrally organized, the old model asks how such a structure assembled so quickly. But UED asks whether the field architecture became visible as a complete coherence pattern. The galaxy may not need to wait for a long mechanical childhood if its structure is an expression of field alignment.

The central regulator, in this reading, does not need to grow separately over billions of years.

It may arise with the galactic field architecture.

The arms, luminosity, rotation, density, and core may be different expressions of one field-realignment event. The outward structure and inward regulator appear together because they are not separate objects; they are complementary conditions of the same system.

This is why the galaxy must be understood as architecture, not accumulation.

Accumulation explains parts.

Architecture explains relation.

Accumulation explains material presence.

Architecture explains coherence.

Accumulation asks how much matter gathered.

Architecture asks how the field became organized.

A galaxy is not a cosmic heap.

It is a structured consequence.

It is the field becoming visible as vast luminous order, while its center remains dark because the same field has crossed into regulatory saturation.

The galaxy shines as architecture.

The center holds as regulation.

Together, they form one field reality.

7.4 — Central Ripple Saturation Zone as Regulator

The center of a galaxy is not merely a location.

It is a condition.

This distinction is important because ordinary thinking treats the center as a point inside a structure. We imagine the galaxy first: stars, gas, dust, arms, rotation, light. Then we imagine a dark object sitting at the middle. The center becomes a place occupied by something. In mainstream language, that "something" is usually called a supermassive black hole.

But under Universal Energy Dynamics, the center is not simply a place where an object sits.

It is the region where the galactic field reaches its deepest regulatory intensity.

This is why the phrase **Central Ripple Saturation Zone** becomes necessary.

A Central Ripple Saturation Zone is not a hole. It is not an infinite-density monster. It is not a cosmic drain. It is a field condition where ripple coherence reaches saturation, ordinary signal recovery fails, and energy-gradient intensity becomes concentrated. From the outside, it appears dark, compact, powerful, and unreachable. But its darkness does not mean emptiness. Its compactness does not automatically mean infinite density. Its no-return behavior does not automatically mean destruction.

It may mean regulation.

To understand this, we must remember what a galaxy is under UED. A galaxy is not merely a pile of stars. It is a large-scale field architecture. It is Structured Vacuum Energy expressing itself through ripple condensation, energy gradients, rotational coherence, luminosity, and equilibrium-seeking behavior. Such a vast structure cannot remain coherent without an inward balancing condition.

The Central Ripple Saturation Zone is that inward condition.

It is where excess field tension concentrates.

It is where ripple behavior reaches a boundary.

It is where the field's outward expression meets its inward limit.

It is where the visible galaxy has its invisible anchor.

In the old imagination, the galactic center consumes. In the UED interpretation, the galactic center regulates. This does not mean it is gentle or passive. Regulation at cosmic scale can be intense. A regulator can absorb, redirect, concentrate, limit, and stabilize. It can appear violent from the outside because the energies involved are enormous. But intensity should not be confused with meaninglessness.

A heart beats with force, but it sustains circulation.

A storm's eye appears strange, but it belongs to the storm's organization.

A nucleus holds extreme concentration, but it gives the atom identity.

Similarly, a Central Ripple Saturation Zone may carry extreme intensity because it performs a structural function. It is not merely a dead endpoint. It may be the boundary through which the galaxy manages its own imbalance.

This is where mainstream black hole language becomes limiting.

When we call the central region a black hole, the mind immediately imagines collapse. Matter falling. Light trapped. A dark pit. A terminal object. The language moves toward destruction. But if we call it a Central Ripple Saturation Zone, the interpretation changes. Now we ask what the region does for the whole galactic field. We ask how it participates in coherence. We ask how it manages gradients. We ask how it relates to rotation, luminosity, structure, and stability.

A better name creates better questions.

The central zone may stabilize the larger galactic architecture by concentrating the strongest field-gradient conditions at the core. Around this core, visible structures can maintain patterned motion. Stars, gas, dust, and energetic emissions are not simply moving around an isolated monster. They are participating in a large-scale field relationship. The dark center provides an inward reference, a saturation boundary, a regulatory threshold around which the visible galaxy expresses structure.

This is why the center may appear so powerful without needing to be interpreted as an impossible singularity.

Its power comes from field concentration.

Its darkness comes from coherence failure.

Its influence comes from gradient intensity.

Its regulatory role comes from its position within the whole field architecture.

In UED, gravity itself is energy-gradient equilibrium. Therefore, the central region of a galaxy is not powerful merely because it contains an impossible amount of matter compressed into an impossible point. It is powerful because the field-gradient condition there is extreme. The region marks the highest saturation intensity of the galactic field.

From this core, the larger galaxy receives its inward balance.

This also explains why such central regions are often connected with energetic phenomena. Around black-hole-like regions, mainstream astrophysics observes accretion-like behavior, emissions, jets, radiation, and intense motion. Under UED, these are not signs of a monster feeding in a crude sense. They may be signs of field redistribution near a saturation boundary. Matter-form and energy patterns approaching the central zone encounter extreme gradient conditions. Some coherence may fail inward. Some energy may redistribute outward. Some field behavior may appear as jets or emissions.

The system is not merely eating.

It is processing imbalance.

This is a major conceptual shift.

A galaxy is full of motion and disturbance. Stars form and die. Gas moves. Radiation spreads. Structural regions interact. The galactic field cannot remain coherent if all disturbances simply accumulate without regulation. The central saturation zone may serve as a field-processing boundary, absorbing or containing excessive ripple disturbance while allowing the larger structure to remain organized.

This is why it should be called a regulator.

Not because it consciously controls the galaxy.

Not because it acts like a machine with intention.

But because its structural role is regulatory. It helps define limits. It concentrates gradients. It marks the boundary of recoverable coherence. It supports the larger architecture by holding the inward saturation condition.

The galaxy shines outward because its field can express visible structure.

The center remains dark because its field has crossed into saturation.

Both are necessary.

Without outward ripple condensation, there is no luminous galaxy. Without inward saturation regulation, the coherence of that luminous system may not sustain itself. The visible and the dark are not enemies. They are complementary expressions of the same field reality.

This also weakens the fear associated with black-hole-like centers. The dark center is not simply waiting to destroy the galaxy. If that were its only role, stable galaxies would be difficult to understand. Instead, the central zone appears to coexist with galactic structure over immense scales. This suggests that the center is not merely destructive. It is relational. It belongs to the system.

The center may consume certain forms of coherence, but it may also preserve the larger coherence.

That is regulation.

A forest fire can destroy parts but reset conditions. A pressure valve releases excess to protect a system. A drain in a basin prevents overflow. These analogies are limited, but they reveal one principle: what appears destructive locally may be regulatory systemically.

The Central Ripple Saturation Zone may operate through such cosmic logic.

It may be the zone where the field says: beyond this point, ordinary ripple structure cannot continue.

It may be the boundary where excess gradient is absorbed.

It may be the inward regulator of galactic identity.

This changes the meaning of the galactic center completely.

The center is not an accident.

It is not merely a black object inside a bright structure.

It is the saturation heart of the field architecture.

And once we understand it this way, the galaxy and its core begin to look less like two separate things and more like one continuous system: outward light held by inward darkness, visible structure balanced by invisible saturation, cosmic beauty sustained by a silent regulatory boundary.

7.5 — Simultaneous Birth: Galaxy and Regulator as One Event

The galaxy and its central regulator may not be two separate births.

This is the idea that dissolves much of the confusion.

In the usual interpretation, the galaxy and the black hole are placed on a timeline. First something must happen, then something else must follow. Either the galaxy forms first and later feeds the black hole, or the black hole forms first and later helps gather the galaxy. The entire discussion becomes chronological because the system is divided into two objects.

But Universal Energy Dynamics does not begin with two objects.

It begins with one field event.

If a galaxy is a field architecture, then its visible structure and dark center are not independent entities accidentally brought together. They are complementary expressions of the same field-realignment process. The galaxy is the outward manifestation of ripple condensation, luminosity, rotation, and large-scale coherence. The central regulator is the inward saturation boundary where field intensity reaches its regulatory limit.

One expresses.

One contains.

One shines.

One holds.

Together, they form the galactic system.

This is why the word "simultaneous" matters. It does not necessarily mean every star, arm, dust lane, and detail appears fully finished at one exact instant. That would be too crude. Simultaneous here means that the foundational field architecture emerges as one unified event. The outward structure and inward regulator belong to the same manifestation threshold. They arise from the same field condition, not from two unrelated histories.

The galaxy does not first become a complete luminous city and then later install a central regulator.

The regulator is part of the original architecture.

A whirlpool does not create its center after rotation is complete. The rotation and center arise together as expressions of one flow condition. A cyclone does not first become a storm and then later decide to form an eye. The eye belongs to the field structure of the storm. A magnet does not create its poles after becoming magnetic. The poles are expressions of the same field arrangement.

Similarly, a galaxy may not produce its central regulator later as an accidental result of collapse.

The central regulator may be present as part of the galactic field's organizing logic from the beginning of coherent manifestation.

This changes the meaning of formation.

In the slow mechanical story, formation means assembly through time. Matter gathers. Stars form. A central black hole grows. The system gradually develops. In the UED view, formation means field realignment reaching a structural threshold. When the field condition becomes ready, outward ripple condensation and inward saturation regulation emerge as two sides of one event.

This does not deny later development.

After the foundational field architecture appears, the galaxy may continue to evolve. Stars may form, die, and redistribute energy. Gas may move. Regions may brighten or fade. The central regulator may interact with surrounding material. The visible shape may refine. The system may merge with others. Local details may change.

But the basic galactic identity does not wait for all these later processes.

It is established when the field reaches coherence.

This is why JWST's observations become so important. When mature-looking galaxies and powerful central black-hole-like regions appear at cosmic depth, the old model becomes uncomfortable because it expects long sequential development. It asks how the galaxy assembled and how the black hole grew so quickly.

UED asks a different question.

What if they did not need to grow separately?

What if the central regulator appeared with the galactic field architecture because the structure required it?

Then the problem of early maturity changes. The central dark region is no longer a separate cosmic monster that somehow became massive too fast. It is the inward regulatory condition of the same field event that produced the galaxy's outward structure. Its maturity is not chronological growth alone. It is structural necessity.

This also explains why the galaxy and its center seem deeply related.

In mainstream observation, correlations between galaxies and their central compact regions are often treated as co-evolution. The galaxy affects the central object. The central object affects the galaxy. They grow together across time. This may be true at the surface level. But UED suggests a deeper relation: they do not merely co-evolve; they co-belong.

They are not two strangers influencing each other.

They are two expressions of one architecture.

The visible galaxy is the field's expanded pattern.

The central regulator is the field's saturated boundary.

The galaxy's arms, rotation, luminosity, density, and central darkness all belong to one field geometry. The outward and inward conditions are not enemies. They balance each other. Outward structure without inward regulation may disperse. Inward saturation without outward expression would not appear as a galaxy. Together, they create a coherent cosmic form.

This is why the central regulator should not be imagined as a late addition.

A late addition belongs to a mechanical construction model. A field architecture does not work that way. In a field, relations arise together. The center and the surrounding pattern are mutually defined. A center is meaningful because there is a structure around it. The structure is coherent because it has an inward reference.

The galaxy and regulator are born as relation.

Not merely as objects.

This is a profound shift from object-thinking to field-thinking.

Object-thinking asks, "Which object came first?"

Field-thinking asks, "What condition produced this relational structure?"

Object-thinking creates chicken-and-egg problems.

Field-thinking sees one system expressing complementary functions.

Under UED, the galactic regulator is not the cause of the galaxy in a crude one-way sense, and the galaxy is not the cause of the regulator in a separate chronological sense. Both are consequences of a deeper field realignment. The field reaches a threshold. Ripple condensation expresses outward. Ripple saturation concentrates inward. Between the two, a galaxy becomes visible.

This also changes the meaning of central darkness.

The darkness is not merely the end of light. It is the inward limit of the same field that produces light outwardly. It is where coherence stops returning, while the surrounding structure continues expressing. It is not outside the galaxy's identity. It is part of that identity.

The galaxy shines because the field condenses.

The center darkens because the field saturates.

Both arise from the same energetic condition.

This is why simultaneous birth is not mystical. It is structural. It is the recognition that organized systems often appear through relations, not through isolated pieces joining later. The field does not first manufacture separate parts and then negotiate their unity. The field expresses unity through different roles.

The galaxy and its regulator are such roles.

One role is visible expansion.

The other is invisible regulation.

One role is luminous pattern.

The other is saturation boundary.

One role is outward coherence.

The other is inward threshold.

Together, they are the galaxy.

This view does not remove the need for observation. It invites better observation. It asks us to study galactic centers not merely as objects to be weighed, but as regulatory field conditions. It asks us to study galaxies not merely as star populations, but as integrated field architectures. It asks us to study early supermassive black-hole-like phenomena not only as growth puzzles, but as signs of simultaneous field emergence.

The old question was: who came first?

The deeper question is: why did we separate them?

Perhaps the galaxy and its central regulator were never two histories.

Perhaps they were always one event seen from two sides.

7.6 — JWST and the Problem of Early Central Regulators

JWST has not only challenged the idea of early galaxies.

It has also challenged the idea of early galactic centers.

This is where the problem becomes sharper. A mature-looking galaxy at cosmic depth is already uncomfortable for the standard timeline. But when that galaxy also appears to contain or suggest a powerful central black-hole-like region, the discomfort grows further. Now the model must explain not only how the galaxy formed so quickly, but also how its central regulator became so massive, so active, or so influential within the same limited chronological window.

In the mainstream framework, this becomes a serious growth problem.

A black hole is usually imagined as something that begins from collapse. A massive star dies. A remnant forms. It grows by consuming surrounding matter. It may merge with other black holes. It may accrete gas. Over time, it becomes larger. Given enough time, a supermassive black hole may appear at the center of a galaxy.

But JWST has made the phrase "given enough time" uncomfortable.

If central black-hole-like regions appear in galaxies placed very early in the cosmic timeline, the question becomes unavoidable: where did they get the time to grow?

This is why mainstream cosmology searches for heavy seeds, direct-collapse black holes, unusually massive early stars, rapid accretion, exceptional feeding, and special early conditions. These ideas are attempts to protect the growth pathway. If the central object is too large too early, then perhaps it began larger than expected. If it grew too quickly, perhaps the feeding rate was unusually high. If ordinary stellar collapse cannot explain it, perhaps another formation channel must be introduced.

These are serious scientific efforts.

But they remain inside the same assumption.

They assume the central object is a separate thing that must grow chronologically.

Universal Energy Dynamics asks whether this assumption is necessary.

If the galactic center is not a black hole in the traditional sense, but a Central Ripple Saturation Zone, then its so-called early maturity may not require the same growth story. It may not be an object that began small and became supermassive by feeding. It may be the inward regulatory condition of the galactic field architecture itself.

This changes the question completely.

Instead of asking, "How did such a massive black hole grow so early?" UED asks, "Why are we assuming the central regulator had to grow separately from the galaxy?"

Instead of asking, "What seed created this monster?" UED asks, "What field condition produced this regulatory saturation zone?"

Instead of asking, "How did it consume enough matter in time?" UED asks, "Was it ever primarily a consuming object, or was it a saturation boundary of the field?"

This is where JWST becomes deeply important for UED.

The telescope is showing systems that strain the slow-growth imagination. It is showing structure where infancy was expected. It is showing compactness where looseness was expected. It is showing central activity where delayed development was expected. The old model responds by accelerating growth. UED responds by questioning whether growth is the right category.

A central regulator may not need to grow like a child.

It may arise like a field boundary.

When a galactic field system manifests, outward ripple condensation creates visible structure: stars, gas, luminosity, rotation, and galactic form. At the same time, inward saturation creates the regulatory core: a zone where field coherence reaches its limit, signal recovery fails, and gradient intensity concentrates. The regulator is not a later addition. It is part of the same field-realignment event.

In this view, an early central regulator is not impossible.

It is expected.

If the galaxy exists as a coherent field structure, the regulatory condition must already be present. A galaxy without such a core would be like a whirlpool without a center, a cyclone without an eye, or a rotating system without an inward reference. The center is not a luxury. It is not decoration. It is structural necessity.

This is why the early appearance of supermassive black-hole-like regions may be less shocking under UED.

They are not necessarily ancient monsters that grew too fast.

They may be central saturation zones that appeared with the field architecture of the galaxy itself.

This also changes how we interpret their apparent mass. In mainstream language, mass is often treated as inventory: how much matter is contained in the object. But in UED, mass is a field-density signature. A central regulator may appear massively influential not because it contains impossible matter compressed into an impossible point, but because the field-gradient condition around it is extremely intense. Its measured influence may reflect saturation strength, not merely accumulated material history.

This is a crucial correction.

If we interpret the central zone as an object, we ask how much it has eaten.

If we interpret it as a field condition, we ask how intensely the field has saturated.

The first question creates the early-growth crisis.

The second question opens the UED reinterpretation.

JWST therefore does not merely reveal early black holes. It may reveal early central regulatory conditions of galaxies. These conditions appear problematic only when we force them into the old story of chronological feeding and object growth. But if the galaxy and its regulator are born together as a unified field event, then the presence of a powerful center at cosmic depth becomes natural.

The regulator was not waiting for the galaxy to mature.

It was part of the galaxy's maturity.

This helps us understand why deep structures can appear coherent. A galaxy is not only a visible spread of matter. It is an integrated field system. Its center, rotation, luminosity, density, and outer structure belong together. If JWST detects compact deep galaxies with signs of central activity, perhaps it is not seeing separate impossible objects. Perhaps it is seeing complete field architectures at deeper observational layers.

This is exactly where the cosmic timeline begins to weaken further.

The old timeline expects sequence: first simple matter, then early stars, then small galaxies, then growing centers, then mature galaxies. But UED expects relation: outward condensation and inward regulation arising together when the field condition crosses threshold. The old timeline sees early central regulators as anomalies. UED sees them as signs that galaxies are field systems, not slow mechanical piles.

The difference is not small.

It changes the entire meaning of cosmic maturity.

A galaxy does not become mature only because enough time has passed. It becomes mature when its field architecture becomes coherent. If a central regulator is already present, that may not mean the object cheated time. It may mean the architecture is already established. The galaxy has a visible body and an invisible regulatory core. It is structurally complete at the field level, even if many internal processes continue evolving.

This is why the JWST problem is not merely a problem for black hole growth.

It is a problem for object-based cosmology.

It exposes the weakness of treating galaxies and their centers as separate things. It exposes the weakness of assuming central regions must grow only by feeding. It exposes the weakness of believing that structure must always follow a slow mechanical ladder.

UED offers another possibility.

The galaxy and its regulator may arise together.

The luminous structure and the dark core may be one event.

The apparent early supermassive black hole may not be early, may not be a monster, and may not be a black hole in the inherited sense.

It may be the central regulatory saturation zone of a deep field architecture.

JWST may not be showing us cosmic impossibility.

It may be showing us that the universe was never building galaxies without their centers.

It may be showing us that the regulator is not a late product of the galaxy.

It is the condition through which the galaxy remains a galaxy.

Closing Reflection — The Darkness That Holds the Pattern

A galaxy does not shine by light alone.

This may sound strange at first, because when we look at a galaxy, light is the first thing that reaches us. We see stars. We see arms. We see dust lanes glowing against darkness. We see structure spread across space like a luminous signature of cosmic order. Naturally, the mind gives importance to what shines.

But the visible is not always the whole truth.

Sometimes the most important part of a structure is not what announces itself loudly, but what quietly holds the structure together. Sometimes the center does not need to shine because its role is not display. Its role is regulation. Its role is balance. Its role is to define the inward condition around which the outward pattern remains coherent.

This is how we must begin to understand the galactic center.

The dark center of a galaxy may not be a cosmic accident. It may not be a late product of collapse. It may not be a monster hiding inside beauty. It may not be an object that grew separately and then occupied the middle of an already formed structure.

It may be the regulatory condition of the galaxy itself.

This chapter has tried to shift the imagination from object to field. In the old view, the galaxy and the black hole are two separate entities. One shines. One is dark. One spreads outward. One sits at the center. One is visible structure. One is mysterious collapse. Once separated in this way, the mind begins asking which came first, which grew faster, which influenced the other, and how the central object became so massive so early.

But perhaps this entire difficulty is created by separation.

Under Universal Energy Dynamics, a galaxy is a field architecture. Its stars, gas, arms, rotation, luminosity, gradients, and center are not merely separate things gathered together. They are expressions of one deeper field condition. The visible galaxy is outward ripple condensation. The dark center is inward saturation regulation. The galaxy's light and the center's darkness are not enemies. They are complementary expressions of one structured event.

The galaxy shines outward because the field condenses into visible form.

The center darkens inward because the field reaches saturation beyond recoverable coherence.

Together, they create the pattern.

This is why the phrase "Central Ripple Saturation Zone" is more meaningful than the inherited image of a black hole. A black hole sounds like an ending. A saturation zone sounds like a condition. A black hole sounds like destruction. A regulator sounds like function. A black hole makes the center appear separate from the galaxy. A central regulator places the center back inside the architecture of the whole.

That correction matters deeply.

If the central dark region is a regulator, then the galaxy does not need to create it later like an afterthought. It arises with the galaxy because the field structure requires inward balance. A whirlpool does not install its center after rotation begins. A cyclone does not add its eye after becoming a storm. A galaxy may not develop its regulator as a separate historical accident. The regulator may belong to the first coherence of the galactic field.

This changes how we understand so-called early supermassive black holes.

The old model asks how they grew so fast.

UED asks whether they needed to grow separately at all.

The old model imagines feeding, accretion, heavy seeds, and rapid chronological growth.

UED imagines field realignment, outward condensation, inward saturation, and simultaneous structural emergence.

This does not deny later evolution. Galaxies can evolve. Central regions can interact with surrounding matter. Energy can redistribute. Stars can form and die. Structures can merge. Nothing in UED requires the galaxy to remain frozen after manifestation. But the foundational relationship between galaxy and regulator may be established from the beginning of coherent field expression.

The galaxy and the regulator may not be two histories.

They may be one system seen from two directions.

This is the real landing of the chapter.

The visible structure and invisible regulator belong together. The outer light and inner darkness form a single field reality. The galaxy is not merely a luminous body with a dark secret at its center. It is a structured field whose darkness may be necessary for its light to remain patterned.

This may also be the deeper lesson of cosmic architecture.

Not everything that sustains a structure is visible.

Not everything dark is destructive.

Not every silence is emptiness.

Not every boundary is an ending.

The dark center may be where the galaxy's imbalance is managed. It may be where excess field tension reaches a saturation condition. It may be where ordinary ripple coherence fails to return, not because reality disappears, but because the field has crossed into regulatory intensity. It may be the inward anchor that allows the outward system to remain coherent.

In that sense, the darkness is not outside the galaxy's beauty.

It is part of its beauty.

The galaxy may shine because its darkness knows how to hold the pattern.

This prepares us for the next movement of the book. We have reinterpreted matter. We have reinterpreted gravity. We have reinterpreted black holes. We have now reinterpreted the galactic center. But one more major assumption remains before this part can complete its work: the way we read redshift, distance, and the illusion of looking back.

Because if galaxies are field architectures and their centers are regulatory saturation zones, then the signals reaching us from cosmic depth must also be read differently. Redshift may not be merely a timestamp.

Distance may not be merely history. Cosmic depth may not be merely the past.

The galaxy has shown us the pattern.

Now the signal itself must be questioned.

Chapter 8 — Redshift, Distance, and the Illusion of Looking Back

A UED reinterpretation of redshift as field strain, ripple tension, and resistance across layers of the present universe.

The Color of a Misread Universe

Redshift is one of the most trusted colors in cosmology.

A signal arrives stretched toward red, and immediately the universe begins to speak in the language of distance, expansion, age, and history. The wavelength has shifted. The object must be far. The light must be old. The structure must belong to the early universe. The farther the redshift, the deeper the past.

This chain has become so familiar that it almost feels automatic.

But every automatic interpretation deserves suspicion.

The observation is real. The shift is real. The signal has changed. The light reaching our instruments does not carry the same wavelength condition we would expect from a nearby source. No serious reinterpretation needs to deny this. The question is not whether redshift exists.

The question is what redshift means.

Modern cosmology has turned redshift into a cosmic timestamp. It has become one of the main tools through which the universe is arranged into a grand timeline. Low redshift belongs to the nearer and more recent universe. High redshift belongs to the distant and early universe. Once this arrangement is accepted, every high-redshift object is immediately placed inside the childhood of cosmic history.

Then the judgment begins.

If a high-redshift galaxy appears faint and primitive, the timeline feels safe. If it appears bright, compact, massive, or organized, the timeline becomes uncomfortable. If a central black-hole-like regulator appears there, the discomfort grows. If many such structures appear, the language of anomaly enters the room.

But perhaps the anomaly is not born from redshift.

Perhaps it is born from the way redshift has been translated.

A measurement becomes dangerous when it is treated as a complete story. Redshift measures a change in the signal. But from that change, we construct distance, age, expansion history, cosmic development, and structural expectation. We convert a shifted wavelength into a chapter of universal biography. Then we force the observed object to behave according to that chapter.

This is where the universe may be misread.

Universal Energy Dynamics asks us to widen the meaning of redshift. It does not treat redshift only as the shadow of distance or the fingerprint of expansion. It asks whether redshift may also carry the signature of field strain, ripple tension, resistance, and alignment delay within Structured Vacuum Energy.

In other words, redshift may not merely tell us how far back we are looking.

It may tell us how deeply the field has been strained.

This is a decisive shift.

If the universe is imagined as empty space expanding like a fabric, redshift naturally becomes evidence of stretching across distance. But if the vacuum is not empty, if it is Structured Vacuum Energy, then a signal does not travel through nothing. It emerges, propagates, distorts, resists, and resolves through field conditions. The signal is shaped not only by separation, but also by the medium's state.

A signal from a high-tension field region may carry more than age.

It may carry resistance.

It may carry compactness.

It may carry energy-gradient history.

It may carry ripple strain.

It may carry the difficulty of resolving through layered field conditions.

This means high redshift may not be a simple label saying, "This belongs to the primitive past." It may be a more complex signature saying, "This signal has emerged through a field condition very different from your local observational environment."

That changes everything.

The so-called early galaxy may not be merely early.

The Little Red Dot may not be merely distant.

The central regulator may not be merely too massive too soon.

They may all be intense field configurations being interpreted through a narrow redshift-to-history pipeline.

This does not mean distance disappears. It does not mean light-travel delay becomes meaningless. It does not mean every cosmological calculation must be thrown away casually. But it does mean that redshift should not be imprisoned inside one interpretation. A shifted signal may carry multiple layers of meaning. To reduce it only to expansion and lookback time may be scientifically convenient, but convenience is not always truth.

A color can be real and still be misunderstood.

A red signal may not only be the mark of a receding object. It may be the visible consequence of strained field interaction. It may be the color of resistance. It may be the universe showing that the signal has passed through a deeper condition than our model has learned to read.

This is why JWST becomes so important again.

JWST is not merely seeing farther. It is receiving signals from field layers that older instruments could not resolve. When those signals are immediately converted into ancient history, the observed structures appear to violate the timeline. But if those signals are also read as field-condition signatures, then the same structures may begin to look less anomalous.

The universe may not be sending us photographs from a dead past.

It may be sending us consequences from a layered present.

Redshift, then, is not the universe's memory in a simple sense. It may be the field's resistance becoming visible through color. It may be the stretched language of a signal struggling through Structured Vacuum Energy. It may be the trace of alignment delay across the continuous Now.

This chapter begins from that possibility.

We are not rejecting redshift.

We are rescuing it from overconfidence.

Because redshift may be one of the greatest clues in cosmology, but only if we stop treating it as a single-purpose clock. The universe may not be saying, "Look how old I am."

It may be saying, "Look how deeply strained my field is."

And perhaps, in that red color, we are not seeing only distance.

We are seeing the hidden tension of reality itself.

8.1 — How Redshift Became a Cosmic Timestamp

Redshift began as an observation.

A signal arrived with its wavelength shifted toward the red end of the spectrum. Light that was expected to appear in one condition appeared stretched into another. This was measurable. It was not imagination. It was not philosophy. It was a physical change detected through instruments.

But what happened next is more important.

The measurement became a meaning.

The meaning became a model.

The model became a timeline.

And the timeline became a worldview.

In modern cosmology, redshift is commonly understood through expansion. If the universe is expanding, then light travelling across expanding space gets stretched. The farther the source, the more the light stretches. The more it stretches, the greater the redshift. This redshift then becomes a marker of distance. Distance becomes lookback time. Lookback time becomes cosmic history.

This chain is now repeated so often that it almost feels like direct reality.

Redshift means distance.

Distance means past.

Past means earlier universe.

Earlier universe means younger structure.

Younger structure means less maturity.

This is how a shifted wavelength becomes a cosmic timestamp.

At first, the process appears reasonable. Light does take time to reach us. Distant signals do not arrive instantly. A signal from a far galaxy carries delay. No one needs to deny this. The issue is not the existence of delay. The issue is the confidence with which delay is converted into complete historical meaning.

Once redshift becomes a timestamp, it begins to control interpretation.

A high-redshift galaxy is no longer merely a galaxy with a strongly shifted signal. It becomes an early galaxy. It is placed inside a young chapter of the universe. Then it is expected to behave according to that chapter. It should be faint. It should be small. It should be rough. It should still be assembling. It should not look too organized, too bright, too massive, or too compact.

If it does, the galaxy becomes a problem.

But the problem did not begin with the galaxy.

The problem began with the timestamp.

This is the subtle danger. Redshift is an observational fact, but the timeline built from it is an interpretation. The fact is that the signal has shifted. The interpretation is that this shift tells us not only distance, but also historical stage, developmental age, structural expectation, and cosmic maturity.

That is a very heavy load to place on one measurement.

A number becomes a chapter.

A wavelength becomes a biography.

A color becomes a clock.

This does not mean redshift is useless. It is extremely useful. It provides important information about signals, motion, distance relations, and cosmic-scale behavior. But usefulness should not be confused with completeness. A useful tool becomes dangerous when it is treated as a final judge.

In the standard cosmic story, redshift became one of the main organizing tools because it helped arrange the universe into a timeline. Low redshift meant nearer and more recent. High redshift meant farther and older. Extremely high redshift meant closer to the beginning. The universe became readable through redshift layers, like pages in a cosmic history book.

This gave cosmology great confidence.

It also created a hidden rigidity.

Because once the universe is arranged in this way, every object receives an assigned historical identity before its full physical condition is understood. A high-redshift structure is not first allowed to be a field condition. It is first placed in the early universe. Then all its features are interpreted through that placement.

This is why JWST caused discomfort.

JWST did not merely detect high-redshift signals. It detected structures whose assigned historical position created contradiction. Their redshift placed them in the early universe, but their brightness, compactness, and apparent maturity seemed to belong to a later stage. The timeline said childhood. The object showed coherence. The model expected infancy. The signal suggested intensity.

So the word anomaly appeared.

But perhaps the anomaly is not in the redshifted object.

Perhaps it is in the way redshift became a timestamp too quickly.

Universal Energy Dynamics does not reject redshift. It reopens it.

UED asks whether redshift may carry more than expansion-based distance and lookback time. If the vacuum is Structured Vacuum Energy, then a signal does not travel through empty nothingness. It passes through field conditions. It carries the consequences of resistance, strain, ripple tension, energy redistribution, and alignment delay. The signal is not merely stretched by distance; it may be shaped by the field through which it resolves.

This widens the meaning of redshift.

A high-redshift signal may not only say, "This object is far and old."

It may also say, "This signal has emerged through a high-resistance field condition."

It may say, "The field state involved is different from the local observational field."

It may say, "The ripple has experienced strain."

It may say, "The energy configuration is compact, tense, or deeply layered."

This does not cancel the older interpretation. It completes what the older interpretation may have oversimplified.

The danger of redshift as a timestamp is that it turns a complex field signal into a simple historical label. It makes us believe we are looking at the universe's childhood whenever we see high redshift. Then, when the so-called childhood looks mature, we blame the universe for being strange.

But perhaps the universe is not strange.

Perhaps redshift is not only a clock.

Perhaps it is also a stress mark.

Perhaps it is not only the signature of distance.

Perhaps it is also the color of field resistance.

This is the shift Chapter 8 must establish. Redshift became the clock of the cosmic story because cosmology needed a way to arrange vast space into readable history. But now JWST has shown that the clock may not be enough. The signal may be carrying more than age. The color may be carrying more than distance. The measurement may be asking for deeper interpretation.

Redshift became powerful because it helped us organize the universe.

Now it must become wiser.

It must help us understand the field, not merely arrange the timeline.

8.2 — The Difference Between Signal Delay and Cosmic History

Light takes time to arrive.

This fact is simple, powerful, and undeniable.

When we see the Moon, we do not see it in the exact same instant in which it exists locally. When we see the Sun, we see it after light has taken several minutes to reach us. When we see distant stars, we see signals that have travelled for years. When we see galaxies at immense distance, the delay becomes enormous.

So, yes, observation involves delay.

But delay is not the same as complete history.

This distinction is essential.

A signal arriving late tells us that transmission has taken time. It tells us that the signal has crossed a certain observational relationship before reaching us. It tells us that what we receive is not locally immediate. But it does not automatically tell us that the object must be interpreted as a primitive historical stage of cosmic development.

That second step is interpretation.

And it is a very large step.

Modern cosmology often moves from signal delay to cosmic history with great confidence. A signal has travelled for a long time; therefore, the object belongs to the past. The object belongs to the past; therefore, it belongs to an early stage of the universe. It belongs to an early stage; therefore, it should be young, immature, faint, and structurally incomplete.

This is where the problem begins.

The first statement is observational.

The later statements are model-based.

The telescope receives a signal. The instrument measures wavelength, brightness, redshift, distribution, and other properties. Then the model places that signal into a cosmic timeline. Once placed there, the object receives a developmental identity. It becomes an early galaxy, an early black hole candidate, an early structure, an early formation problem.

But perhaps we must separate what is seen from what is assigned.

What is seen is signal.

What is assigned is history.

The difference may look small, but it changes everything.

Suppose we receive an old letter. The letter took time to reach us. That is delay. But understanding the person who wrote it requires more than knowing when the letter was sent. We need context, condition, intention, environment, and meaning. If we judge the entire person only by the delay of the letter, we may misunderstand the person completely.

Cosmic signals are similar.

A signal may arrive after a long delay, but the meaning of that signal depends on the field condition from which it emerged, the medium through which it travelled, the resistance it encountered, and the interpretive model through which we read it.

Signal delay is not enough.

It gives timing relation.

It does not alone give complete cosmic identity.

This is where Universal Energy Dynamics asks us to be more careful. Under UED, the universe is not treated as a simple archive of past frames. It is a continuous field reality. What we call light, redshift, distance, and observation are all consequences of energy interacting through Structured Vacuum Energy. A signal does not travel through dead emptiness. It travels through field condition.

Therefore, the arriving signal may carry more than delay.

It may carry field strain.

It may carry ripple tension.

It may carry resistance.

It may carry alignment delay.

It may carry the signature of the field state through which it has resolved.

If we ignore all this and treat delay as history alone, we may reduce a complex field consequence into a simple chronological label.

That reduction creates anomalies.

A high-redshift galaxy is placed into the early universe. Its signal delay is converted into age. Its age is converted into expected immaturity. Then JWST observes brightness, compactness, density, and structure. Suddenly, the galaxy looks too mature. But perhaps the problem is not maturity. Perhaps the problem is the way delay was converted into developmental expectation.

The signal was delayed.

But the object was judged.

This is the danger.

In human life, we understand that delay and history are different. A delayed message does not fully define the sender. A delayed photograph does not fully define the present condition of a place. A fossil does not fully explain a living ecosystem. Yet in cosmology, we often speak as if delayed light gives us a direct and complete view of cosmic childhood.

It gives us a relation.

It gives us evidence.

It gives us a signal.

But the complete historical meaning is constructed.

This does not make cosmology useless. It makes cosmology responsible. Science must build models; otherwise, data remains scattered. But science must also remember which parts are measured and which parts are inferred. Redshift is measured. Signal delay is calculated. Cosmic age is inferred through a model. Developmental expectation is added through theory.

When these layers are mixed together too quickly, confidence becomes dangerous.

UED does not deny that distant signals involve lookback relation. It questions whether lookback relation should dominate the entire meaning of observation. A signal may be delayed, but it may also be shaped by the present structure of the field. The universe may not be showing us dead historical scenes. It may be presenting field consequences that our mind arranges as history.

The past is meaningful to the observer.

But the field may be operating in a continuous present.

This is why JWST is so disruptive. It receives signals that mainstream interpretation places very far back in time. But the structures revealed by those signals do not always behave according to the assigned historical stage. This mismatch tells us that signal delay alone may not be enough to define cosmic maturity.

The question must change.

Instead of asking only, "How old is this light?"

We must also ask, "What field condition shaped this signal?"

Instead of asking only, "How far back are we looking?"

We must also ask, "How deeply into the field are we resolving?"

Instead of asking only, "Why does this object appear too mature for its time?"

We must ask, "Was time the right measure of its maturity?"

This is the essential difference between signal delay and cosmic history.

Signal delay is physics.

Cosmic history is interpretation built upon physics.

The first must be respected.

The second must remain open to revision.

Redshift and delay may tell us that the signal has crossed a vast observational relationship. But they may not tell us, by themselves, that the universe was young, primitive, or incomplete in the way our timeline expects. That expectation belongs to the model.

And if the model repeatedly produces anomalies, then the model must be examined.

Perhaps JWST is not showing us impossible structures from the ancient past.

Perhaps it is showing us that the ancient past is not as simple as our interpretation made it.

Or more radically, perhaps it is showing us that cosmic depth is not merely pastness at all.

It may be field depth.

It may be resistance.

It may be structured presence arriving through delay.

The signal arrives late.

But the meaning must not arrive prematurely.

8.3 — Field Strain: Redshift as Resistance Signature

Redshift may be more than stretched light.

It may be strained light.

This distinction is central to the UED reinterpretation.

In the standard cosmological reading, redshift is usually connected to expansion and distance. The universe expands, light stretches, wavelength increases, and the signal shifts toward red. This explanation has become deeply embedded in modern cosmology. It allows redshift to become a distance marker, and then a time marker, and then a historical marker.

But Universal Energy Dynamics asks us to widen the meaning.

What if the signal has not merely stretched because of cosmic expansion?

What if the signal has also carried the effect of field resistance?

What if the redshift is telling us not only how far the signal has travelled, but what kind of field condition it has passed through or emerged from?

This is where redshift becomes a resistance signature.

In UED, the vacuum is not empty. It is Structured Vacuum Energy. It is not a dead background through which light moves freely without consequence. It is the deeper field condition within which energy behaves, ripples propagate, matter condenses, gravity emerges, and signals resolve into observation.

Therefore, a signal cannot be treated as if it travels through nothing.

It travels through condition.

It travels through field state.

It travels through resistance, strain, gradient, tension, and alignment delay.

If the field were absolutely empty, perhaps the signal could be imagined as a simple messenger crossing a passive gap. But if the vacuum is structured, then every signal carries the memory of interaction. Not memory in the human sense, but consequence. The signal becomes shaped by the medium through which it resolves.

This is why redshift must be reconsidered.

A shifted signal may not only indicate that space has stretched. It may indicate that the signal has moved through a field condition where energy redistribution, ripple resistance, and structural tension have altered its observable state. The red color may be the visible result of the signal losing coherence, adjusting wavelength, or resolving differently because of the resistance profile of the field.

In simple words, redshift may be the color of field strain.

This does not mean distance is irrelevant. It does not mean expansion-based interpretations have no value. It means they may not be complete. A single observed effect can arise through multiple interacting conditions. If we reduce redshift only to distance and expansion, we may miss the deeper field behavior hidden inside it.

Consider sound moving through different environments. Sound does not behave the same way in open air, underwater, inside a hall, or through a dense wall. The signal changes because the medium matters. It may weaken, distort, echo, delay, or shift in quality. The signal is not just a message from the source; it also carries the effect of the path and medium.

Light and cosmic signals may deserve similar humility.

If the cosmic field is structured, then redshift may be telling us about the state of that structure.

A high-redshift signal may not simply say, "I came from very far away."

It may also say, "I came through a field condition of high resistance."

It may say, "I emerged from a compact energy configuration."

It may say, "The ripple state was under strain."

It may say, "The alignment delay was intense."

It may say, "The field through which I resolved is not like your local stabilized field."

This opens a much richer interpretation of JWST observations.

When JWST sees high-redshift galaxies that appear bright, compact, mature, or centrally organized, the standard timeline becomes uncomfortable because high redshift places those structures into a young cosmic stage. But if high redshift also indicates field strain, then those structures may not be young in the simple chronological sense. They may be high-tension regions of the present universe whose signals arrive through strong resistance conditions.

Their redshift may not be only a timestamp.

It may be a stress signature.

This changes the meaning of "far."

Far may not be only spatially distant. It may also be field-different. A signal from such a region may be difficult to resolve not merely because it has travelled long, but because the field state involved is highly strained compared to our local observational condition.

In this view, redshift becomes a diagnostic tool for field behavior.

A galaxy with high redshift may be understood as a region where ripple conditions differ significantly from our local field. The light does not simply arrive as a delayed photograph. It arrives as a transformed consequence of field interaction. Its redness is not only telling us where it came from in the cosmic map. It may be telling us how the field resisted, stretched, delayed, or reshaped the signal.

This interpretation also helps explain why some high-redshift structures appear unexpectedly intense.

If the field condition is strained, then energetic expression may be strong. If ripple tension is high, luminosity may appear extreme. If matter is ripple condensation within SVE, then compact structures may represent concentrated field states. If central regulators arise as saturation zones, then their early appearance may not indicate impossible growth but intense field architecture.

The signal carries this condition.

Redshift becomes the visible clue.

This is why the standard interpretation may create artificial anomalies. It reads redshift mainly as historical distance. Then it places the object into an early stage. Then it expects immaturity. But if redshift also carries field strain, then the object's apparent maturity is not necessarily contradictory. It may be exactly what a high-strain field condition should produce.

A strained field may create compact structure.

A resistant field may delay alignment.

A saturated field may produce redshifted signals with unusual intensity.

A high-tension region may look anomalous only when interpreted as a primitive historical stage.

Under UED, redshift is therefore not merely a passive effect.

It is active evidence.

It is evidence that the signal has been shaped by the field. It is evidence that cosmic observation is not simple storytelling across time. It is evidence that the vacuum participates. It is evidence that the universe is not merely a collection of objects separated by empty distance, but a continuous structured field where every signal carries the consequence of its passage and origin.

This also means redshift should not be treated as a single-purpose measurement.

It should be read in layers.

First, as wavelength shift.

Second, as observational relation.

Third, as possible distance marker.

Fourth, as possible alignment delay.

Fifth, as field strain signature.

Sixth, as ripple resistance evidence.

Only then does the interpretation become mature enough.

A measurement becomes dangerous when it is asked to say too little and too much at the same time. In modern cosmology, redshift is often asked to say too much about history while saying too little about field condition. It becomes the clock of the universe but not the pressure gauge of the field.

UED restores the pressure gauge.

It asks us to read redshift as the trace of resistance. It asks us to treat the red color not merely as ancient distance, but as the field's visible strain. It asks us to examine whether highly redshifted signals are revealing layers of Structured Vacuum Energy under different conditions of density, tension, and coherence.

This has major consequences.

If redshift is field strain, then high-redshift objects are not automatically primitive. If redshift is resistance signature, then brightness and compactness may belong to the same condition producing the shift. If redshift includes alignment delay, then the Hubble tension may not be a crisis of expansion alone, but a sign of state-dependent resolution delay.

This chapter will later lead us there.

For now, the key correction is clear.

Redshift is not only the universe stretching light across distance.

It may be the field marking the signal with strain.

And if we learn to read that strain properly, many cosmic anomalies may begin to look less like impossible objects and more like misunderstood field conditions.

8.4 — Ripple Tension Across the Continuous Present

A signal is not a fossil.

This may be difficult to accept because we have been trained to imagine cosmic light as ancient evidence preserved across space. A distant signal is treated like an old photograph, a relic from a dead moment, a piece of the universe's past arriving late to our instruments. In this imagination, telescopes become archaeological tools, and deep space becomes a museum of ancient light.

But perhaps this picture is incomplete.

Under Universal Energy Dynamics, a signal is not merely a fossil from the past. It is a field consequence moving through the continuous present. It is ripple behavior interacting with Structured Vacuum Energy. It is not simply "old light" travelling through empty space. It is energy expression shaped by field tension, resistance, alignment delay, and resolution conditions across the SVE continuum.

This changes the meaning of observation.

When JWST receives a deep redshifted signal, it is not merely receiving a dead message from a dead cosmic scene. It is resolving a ripple consequence that has passed through layered field conditions and has reached the instrument in a transformed state. The signal carries delay, yes. But it also carries the field's resistance. It carries the difficulty of propagation. It carries the imprint of the medium. It carries the tension of the path.

This is why the phrase "looking back in time" must be handled carefully.

It may be useful as an introductory explanation, but it is not enough for deeper interpretation. If the universe is a structured field, then observation is not only backward-looking. It is also depth-resolving. We are not merely retrieving old scenes. We are interacting with signals whose present arrival contains the consequences of field conditions through which they have resolved.

The signal is arriving now.

The detection is happening now.

The instrument is interacting now.

The interpretation is being made now.

The field consequence is becoming measurable now.

This does not erase transmission delay. It places delay inside a broader framework. The signal may have a travel history, but it is not only history. It is also a present event of resolution. Until it interacts with the detector, it is not an image in our world. It is a field consequence reaching observational completion through interaction.

This is why UED speaks of the continuous present.

The universe is not a library of dead frames waiting for us to open them. It is an active field reality in which consequences are continuously resolving. What we call past may be the mind's arrangement of received consequences. What we call distant history may be a model-based translation of delayed signal reception. But the field itself is not necessarily preserving itself as a human-style archive.

The field is active.

It strains.

It resists.

It aligns.

It saturates.

It resolves.

It expresses.

A redshifted signal, therefore, is not merely a label from yesterday's universe. It is a ripple relation across the continuous Now. It tells us not only that something is far according to our model, but that the signal has undergone field interaction. It has crossed resistance layers. It has carried tension. It has arrived in a changed state because the field through which it moved is not empty.

This is where ripple tension becomes important.

A ripple carries pattern only if the medium allows it. If the medium is calm and stable, the ripple may remain coherent. If the medium is disturbed, strained, dense, or resistant, the ripple changes. It may stretch, weaken, distort, delay, or lose part of its coherence. The final signal carries not only the source condition, but also the field condition.

This is true in ordinary systems. A sound travelling through a quiet room is different from a sound travelling through a storm. A wave moving across calm water is different from a wave moving through turbulent flow. A message transmitted through a clean channel is different from one transmitted through interference. The message is still real, but it is not independent of the medium.

The same principle must be applied cosmically.

If Structured Vacuum Energy is the medium of manifestation, then no signal is medium-free. Cosmic light does not travel through nothing. It travels through the field. It resolves through gradients, resistance, density, and tension. The redshifted signal is therefore not a pure historical postcard. It is a present consequence carrying the marks of ripple tension.

This helps us reinterpret high-redshift structures.

A galaxy observed at high redshift may not simply be an immature object from an ancient stage. It may be a field structure whose signal has emerged through intense ripple tension. Its brightness, compactness, redness, and apparent maturity may belong together. They may all be signatures of a high-strain field condition rather than disconnected puzzles.

The old interpretation separates them.

Redshift becomes distance.

Distance becomes past.

Brightness becomes excessive.

Compactness becomes strange.

Maturity becomes impossible.

UED reunites them.

Redshift becomes field strain.

Brightness becomes energetic expression.

Compactness becomes local condensation.

Maturity becomes coherence.

The entire observation becomes one field consequence.

This is why the continuous present matters. If we treat high-redshift signals only as old light, we automatically place the observed structure into a developmental past. Then we ask why it looks too mature for that past. But if we treat the signal as ripple behavior arriving through layered present-field conditions, then its apparent maturity may not violate anything. It may reveal the intensity of the field layer being resolved.

JWST, in this sense, is not merely seeing old objects.

It is resolving deeper ripple conditions.

It is not only looking at what the universe was.

It is revealing how deeply the universe exists.

This does not mean the instrument changes the universe. It means the instrument crosses earlier observational limitations. Older telescopes could not resolve certain field layers with the same sensitivity. JWST can receive signals shifted into infrared, detect faint structures, and expose regions whose ripple consequences were previously beyond our observational reach. This improved sensitivity does not only extend distance. It extends field penetration.

The telescope becomes a participant in resolution.

Not in the sense of creating the object, but in the sense of completing the observational relation. The signal becomes data only when the field consequence reaches the detector and is interpreted. That interpretation must now become more mature. It must not immediately reduce every deep signal to a dead past. It must ask what ripple tension, field strain, and alignment delay are being carried by the signal.

This also changes the emotional relationship with cosmic time.

We are not simply late spectators watching an old film.

We are observers inside the same continuous field, receiving consequences from different layers of field condition. The difference between local and distant is not only a clock difference. It may also be a field-state difference. The distant signal may represent a region of the universe whose energy configuration, resistance profile, and ripple behavior differ sharply from our local stabilized field.

This is why redshift may be so powerful.

It may be the visible signature of that difference.

It may be the signal's way of showing how much tension it has carried.

It may be the field's way of saying: this layer is not like yours.

Once this is understood, the illusion of looking back begins to weaken. We are not simply looking backward into a cosmic childhood. We are looking across field depth, through resistance, into layers of structured energy that our old timeline may have misnamed as ancient infancy.

The signal is not a fossil.

It is a consequence.

It is not merely old.

It is strained.

It is not merely distant.

It is field-marked.

It is not merely historical.

It is present at the moment of observation.

That is the deeper meaning of ripple tension across the continuous present. The universe may not be sending us dead memories. It may be revealing active field conditions through signals that arrive after delay but become meaningful only now.

And if we learn to read those signals as field consequences, the so-called anomalies of cosmic depth may begin to speak a different language.

8.5 — Distance as Field Depth, Not Only Spatial Separation

Distance is one of the easiest ideas to misunderstand because it feels so obvious.

Something is near.

Something is far.

A village is farther than a house. A city is farther than a village. The Moon is farther than the clouds. The Sun is farther than the Moon. A galaxy is farther than a star. The mind understands distance as separation. It imagines space as an interval between objects, a measurable gap across which light travels.

This is useful.

But cosmic distance may not be only a gap.

It may also be depth.

Not merely depth in space, but depth in field condition.

This is the UED correction.

In ordinary cosmology, distance is strongly tied to spatial separation and lookback time. A far object is interpreted as an old signal. A high-redshift object is placed deeper into the cosmic past. Distance becomes a historical ladder. The farther we look, the earlier the universe is assumed to be.

But if the vacuum is Structured Vacuum Energy, then distance cannot be treated only as empty separation. A signal does not cross a dead gap. It travels through field condition. It moves through resistance, tension, gradient, density differences, and alignment delays. Therefore, cosmic distance may include more than "how far away the object is." It may also include "how different the field condition is from our local observational state."

This distinction is powerful.

Two regions may be separated spatially, but they may also differ structurally. Their field conditions may not be identical. Their energy density, ripple tension, resistance, coherence, saturation, and alignment thresholds may vary. A signal arriving from such a region carries not only the fact of separation, but also the consequences of moving through or emerging from that field condition.

This is what I call field depth.

Field depth means that the universe is not simply spread across space like objects placed on a table. It is layered through conditions. Some regions may be locally stabilized. Some may be highly strained. Some may carry compact energy configurations. Some may exist near saturation boundaries. Some may produce signals that are difficult to resolve because the field resistance is high.

When we observe such regions, we may mistakenly call them merely distant.

But they may be field-deep.

This changes the meaning of cosmic observation.

If distance is only spatial, then high-redshift galaxies are mainly far-away objects from the past. Their maturity becomes a problem because the model assigns them to a young stage of the universe. But if distance also includes field depth, then the same galaxies may be understood as structures from a highly strained layer of the present field. Their redshift, compactness, brightness, and coherence may all be connected to their field-depth condition.

The old model asks: how far back is this object?

UED asks: how deeply strained is the field from which this signal resolves?

The old model asks: how much time did this galaxy have to grow?

UED asks: what field condition allowed this structure to condense?

The old model asks: why does it look mature at such distance?

UED asks: why are we treating field depth as developmental infancy?

This is not a rejection of spatial separation. It is an expansion of interpretation. Objects may still be far in the ordinary sense. Signals may still take time to reach us. But the meaning of "far" becomes richer. Far is not only a number on a cosmic ruler. Far may also indicate a difference in field state.

A simple analogy can help.

If a person walks across a flat road, distance is easy to measure. Ten kilometers means ten kilometers. But if another person walks ten kilometers through a mountain, forest, swamp, and storm, the same spatial distance does not represent the same experience. The path has resistance. The environment changes the journey. The difficulty is not only in the length, but in the condition of the medium.

Similarly, a cosmic signal does not simply cross neutral emptiness. It resolves through field condition. Its final state may reflect not only how far it has travelled, but what kind of field it has encountered.

This is why redshift can become field-depth evidence.

A high redshift may not only mean the signal came from far away. It may mean the signal passed through or emerged from a layer of high resistance. It may mean the field condition was compact, strained, or saturated. It may mean the alignment delay was strong. It may mean the signal had to resolve through a condition very different from our local field environment.

In that case, redshift becomes less like a postal address and more like a pressure mark.

It tells us not only where the signal came from, but how the field behaved.

This also helps reinterpret JWST.

JWST may not be merely extending our view along a longer spatial line. It may be penetrating deeper field layers. Its infrared sensitivity allows it to detect signals that older instruments could not resolve. Those signals may belong to regions where field resistance is high and ordinary visible observation is insufficient. What appears as deep distance may also be deep field condition.

This is why JWST's discoveries seem disruptive. It is not only seeing farther. It is seeing differently. It is receiving signals from regions whose field-depth conditions may not obey the developmental assumptions of the cosmic timeline. If those signals are immediately converted into ancient infancy, the structures become anomalous. But if they are read as field-depth signatures, the anomaly begins to soften.

A compact red object may not be a small premature galaxy.

It may be a high-field-depth condensation.

A bright high-redshift structure may not be an impossible early object.

It may be a strained field region expressing intense energy release.

A central regulator at cosmic depth may not be a black hole that grew too quickly.

It may be the saturation boundary of a field architecture already coherent at that layer.

This is why distance must be reinterpreted.

Distance is not wrong.

But distance alone is not enough.

Spatial separation tells us one part of the story. Signal delay tells us another. Redshift tells us another. Field resistance tells us another. Alignment delay tells us another. Only when these are read together does the observation become more complete.

The mistake is to let distance dominate everything.

Once distance becomes history, and history becomes age, and age becomes expectation, the object loses its right to be interpreted as field condition. It is forced into a timeline before its deeper nature is understood. That is how cosmic anomalies are manufactured.

Field depth restores the missing layer.

It allows us to say: yes, the signal is distant, but it may also be strained. Yes, the light is delayed, but it may also be resistance-marked. Yes, the redshift is high, but it may also indicate a different field state. Yes, the object appears at cosmic depth, but depth is not automatically childhood.

This is a more mature way to read the universe.

The universe may not be arranged merely as near and far.

It may be arranged as layers of field condition.

Some layers are easier to resolve. Some resist. Some saturate. Some condense intensely. Some appear dark. Some appear red. Some appear mature where the timeline expects youth.

Under UED, this is not confusion.

It is field depth becoming visible.

Distance may measure separation, but field depth explains behavior. And if we learn to read both together, the universe no longer appears as a broken timeline. It begins to appear as a structured reality whose signals carry the tension, resistance, and depth of the field itself.

8.6 — Why High Redshift Structures Look Anomalous

High redshift structures look anomalous because we do not allow them to appear first as structures.

We immediately place them inside a story.

A signal arrives strongly redshifted. The model interprets the redshift as great distance. Great distance becomes deep lookback time. Deep lookback time becomes early universe. Early universe becomes youth. Youth becomes immaturity. Then, when the structure appears bright, compact, dense, organized, or centrally active, we call it strange.

But perhaps the strangeness does not begin in the structure.

It begins in the chain of interpretation.

This is why high redshift galaxies, Little Red Dots, and early central regulators appear so disturbing. They are not merely observed; they are assigned a role before they are understood. They are treated as objects from cosmic childhood. Once that role is assigned, they must behave like children of the universe. They must look unfinished. They must look weak. They must look primitive. They must appear as early attempts at structure, not as coherent expressions of structure.

But JWST has not been so obedient.

It has shown us deep structures that carry unexpected intensity. Some appear brighter than expected. Some appear more compact. Some seem too dense or too well organized. Some suggest central activity where the old timeline expects delayed growth. The old interpretation responds by asking how such things could happen so early.

That question sounds scientific.

But it may already be biased.

Because "so early" is not directly observed. It is interpreted through redshift, distance, expansion, and lookback time. The structure is then judged according to that interpretation. The moment we say, "How did this form so early?" we have already accepted that the redshift must be read primarily as chronological infancy.

Universal Energy Dynamics does not accept this automatic step.

Under UED, high redshift may not be only a marker of distance and pastness. It may also be a signature of field strain, ripple tension, resistance, compact energy configuration, and alignment delay within Structured Vacuum Energy. Therefore, a high-redshift structure may not be merely an old photograph from a primitive universe. It may be a signal from a deep field condition.

This changes everything.

A Little Red Dot may not be a tiny early galaxy that somehow became too bright too quickly. It may be a compact field-condensation zone where ripple tension and energy density are extremely high. Its redness may not only be a timestamp. It may be the visible signature of field resistance. Its compactness may not be a problem. It may be evidence of local saturation and intense alignment.

A mature-looking high-redshift galaxy may not be a structure that cheated time. It may be a coherent field architecture whose signal has been interpreted through a narrow chronological lens. Its apparent maturity may not indicate old age. It may indicate field coherence. Its brightness may not only require accelerated star formation. It may also reflect energetic release from field realignment.

A central black-hole-like region at cosmic depth may not be a supermassive monster that grew impossibly fast. It may be a Central Ripple Saturation Zone arising with the galactic field structure itself. Its presence does not necessarily mean the universe built a black hole too quickly. It may mean that the galaxy and its regulator appeared as one field-realignment event.

The anomaly appears because the model separates what the field may hold together.

It separates galaxy and regulator.

It separates redshift and field strain.

It separates distance and field depth.

It separates matter and vacuum.

It separates gravity and energy-gradient behavior.

Then it tries to rebuild the whole universe through a chronological sequence of separated pieces.

UED moves in the opposite direction. It reunites the observation into one field reading. High redshift, brightness, compactness, central saturation, and apparent maturity may not be separate puzzles. They may be connected expressions of the same field condition.

This is why high redshift structures look anomalous under mainstream interpretation but meaningful under UED.

The old model says: high redshift means early, and early should be immature.

UED says: high redshift may indicate field strain, and field strain can produce intense structure.

The old model says: deep objects should be primitive.

UED says: deep field layers may be highly condensed, resistant, and coherent.

The old model says: maturity at high redshift is a problem.

UED says: maturity may mean coherence, not age.

This does not mean every high-redshift object must be explained the same way. Some may be ordinary galaxies. Some may involve intense star formation. Some may involve active central regions. Some mass estimates may change. Some interpretations may be revised as data improves. UED does not need to flatten all observations into one formula.

But it does insist on one thing: do not turn redshift into a rigid historical label too quickly.

Because once that happens, the object is trapped.

It is no longer allowed to be a field condition. It becomes an early object. Then its brightness becomes excessive, its compactness becomes puzzling, its regulator becomes impossible, and its coherence becomes an anomaly.

The problem is not that the universe is producing impossible structures.

The problem is that we are reading field-depth signals as developmental childhood.

JWST may be exposing this mistake. It may be showing that cosmic depth is not always cosmic infancy. It may be showing that high redshift does not always mean primitive structure. It may be showing that the universe is layered through field conditions more complex than the old timeline can absorb.

This is why the anomalies gather around high redshift.

High redshift is where the old interpretation becomes most confident and most vulnerable at the same time. It is confident because the model has already arranged such objects into the early universe. It is vulnerable because sharper observation now shows that the so-called early universe does not always behave like a cosmic nursery.

The higher the redshift, the heavier the interpretation.

And the heavier the interpretation, the more dramatic the anomaly when reality resists it.

Under UED, we do not need to deny the measurement. We need to deepen its reading. Redshift is real, but its meaning may be layered. It may carry distance, delay, strain, resistance, and field-depth information together. If we read only one layer and ignore the others, we create confusion.

High redshift structures look anomalous because we are asking them to fit inside a narrow timeline.

But perhaps they are not timeline failures.

Perhaps they are field signatures.

Perhaps their brightness is not premature.

Perhaps their compactness is not impossible.

Perhaps their central regulators are not too early.

Perhaps their redshift is not merely the mark of ancient distance, but the color of a strained field becoming visible.

The anomaly may arise not from the structure itself, but from turning redshift into a rigid historical label.

Once that label is loosened, the universe begins to speak differently.

Closing Reflection — We May Not Be Looking Back, But Looking Deeper

Perhaps JWST is not only looking backward.

Perhaps it is looking deeper.

This one shift changes the emotional foundation of modern cosmology. For a long time, we have been trained to imagine the universe as a timeline stretched across space. The deeper we look, the earlier we see. The earlier we see, the younger the universe must be. The younger the universe is, the less mature its structures should appear.

This arrangement gave us confidence.

It gave us a cosmic story.

It gave us a beginning, a sequence, and a sense of intellectual control.

But the universe may not be so obedient to our storytelling.

Redshift has carried too much responsibility in this story. It began as a measured shift in wavelength, but slowly became a cosmic clock. It became a distance marker, then an age marker, then a developmental label. Once an object was redshifted enough, it was placed into the early universe. Once placed there, it was judged according to the expectations of earlyness.

That is how anomalies were born.

Not always from the object.

Often from the label.

A galaxy appeared bright, but the timeline expected faintness.

A structure appeared compact, but the timeline expected looseness.

A central regulator appeared powerful, but the timeline expected delayed growth.

A Little Red Dot appeared intense, but the timeline expected primitive formation.

The universe showed structure.

The model saw contradiction.

But under Universal Energy Dynamics, redshift may not be merely the universe's memory. It may be the field's resistance becoming visible. It may be the trace of ripple tension across Structured Vacuum Energy. It may be the color of strain, the mark of alignment delay, the sign that the signal has emerged through a field condition very different from our local observational environment.

This does not mean distance disappears.

This does not mean signal delay is false.

This does not mean all existing redshift interpretation is useless.

It means the meaning is incomplete when treated only as history.

A signal can be delayed and strained.

It can be distant and field-marked.

It can carry lookback relation and resistance signature.

It can arrive from far away and still reveal the condition of the field through which it has resolved.

This is why the phrase "looking back in time" must be handled carefully. It is useful as an introduction, but dangerous as a final interpretation. The universe may not be storing old scenes like a cosmic museum. It may not be handing us ancient photographs from a dead past. It may be presenting field consequences through layered resistance, and our mind may be translating those consequences into historical frames.

JWST, then, becomes more than a telescope of distance.

It becomes an instrument of field penetration.

It receives signals that older instruments could not resolve. It exposes layers of cosmic structure that were previously hidden. It brings into view regions of compactness, brightness, strain, saturation, and coherence that the older timeline was not ready to absorb. When these signals are forced into a simple early-universe frame, they become anomalies. When they are read as field-depth signals, they begin to make deeper sense.

This is the real transition this chapter has prepared.

Redshift is not rejected.

It is reinterpreted.

Distance is not denied.

It is expanded.

The past is not erased.

It is removed from the throne.

The universe may be less like a road behind us and more like a depth before us. What we call far may also be field-deep. What we call old may also be resistant. What we call early may also be high-tension. What we call anomalous may simply be misunderstood field behavior.

This gives us a new way to read cosmic observation.

A red signal is not only a timestamp.

It may be a pressure mark.

A distant structure is not only an ancient object.

It may be a deeper field condition.

A mature high-redshift galaxy is not automatically a contradiction.

It may be coherent field architecture appearing through strained resolution.

This does not make the universe smaller. It makes it larger again. It frees the cosmos from being reduced to a human-style biography. It allows reality to be read not only through before and after, but through condition, resistance, alignment, saturation, and consequence.

The universe may not be saying, "Look how young I once was."

It may be saying, "Look how deeply structured I am."

And perhaps that is why JWST feels so disruptive. It is not merely showing us more. It is forcing us to change the meaning of seeing. It is asking us to stop treating every distant signal as a child of the past and begin reading it as a messenger of field depth.

This chapter closes second part with that realization.

Matter has been reinterpreted as field consequence.

Gravity has been reinterpreted as energy-gradient equilibrium.

Black holes have been reinterpreted as Ripple Saturation Zones.

Galactic centers have been reinterpreted as central regulators.

And now redshift has been reinterpreted as field strain, ripple tension, and resistance across the continuous present.

The old cosmic picture has not been destroyed by argument.

It has been loosened by reinterpretation.

Now we are ready for the next question.

If redshift is not merely the clock of the universe, then what is time?

If cosmic depth is not simply cosmic past, then what does it mean to say that the universe is expanding, aging, or evolving?

If signals arrive through alignment delay, then perhaps time itself must be re-examined.

We may not be looking back.

We may be looking deeper.

And if we are looking deeper, then the universe has never been a dead archive of old light.

It has always been a living field of structured consequence, revealing itself only as far as our instruments, intelligence, and courage are ready to resolve it.

Interlude

When the Signal Stops Being History

There comes a point where reinterpretation is no longer about one object.

It is no longer about one galaxy, one red dot, one black hole, one redshift measurement, or one tension in the data. The issue becomes deeper. The entire way of reading the universe begins to change.

This part of this book has attempted that shift.

Matter was no longer treated as a dead thing placed inside empty space. It was reinterpreted as ripple condensation within Structured Vacuum Energy. Gravity was no longer treated as an invisible pull or merely the bending of an empty spacetime fabric. It was reinterpreted as energy-gradient behavior seeking balance. Black holes were no longer accepted as cosmic monsters of infinite density. They were reinterpreted as Ripple Saturation Zones. Galactic centers were no longer treated as accidental dark objects. They were reinterpreted as central regulators. Redshift was no longer treated only as the clock of cosmic history. It was reinterpreted as field strain, ripple tension, and resistance across layered reality.

Once these shifts are made, the old cosmic picture cannot remain the same.

Because the old picture depends on a simple arrangement.

Matter exists. Gravity pulls. Space expands. Light travels. Redshift measures distance. Distance becomes past. Past becomes early universe. Early universe becomes developmental childhood.

This arrangement gave modern cosmology its movie reel.

The universe began.

The universe expanded.

The universe cooled.

The universe formed particles.

The universe formed stars.

The universe formed galaxies.

The universe matured.

And now, through telescopes, we look backward at earlier frames.

This story is elegant.

It is also deeply human.

The mind loves sequence. It loves before and after. It loves beginning, development, maturity, and conclusion. It understands life through childhood, youth, adulthood, and aging. It understands construction through foundation, walls, roof, and completion. It understands learning through ignorance, effort, knowledge, and mastery.

So when cosmology gave the universe a similar timeline, the mind accepted it easily.

But the universe may not be a human biography.

This is where next part begins.

If matter is field consequence, if gravity is gradient equilibrium, if black holes are saturation zones, if galactic centers are regulators, and if redshift carries field strain, then time itself can no longer sit untouched at the center of interpretation.

Time must be questioned.

Not casually. Not dramatically. Not in the shallow sense of saying time does not exist. Experience clearly involves duration. Events unfold. Signals arrive after delay. Processes require sequencing. Bodies age. Stars change. Instruments measure intervals. No serious framework should deny lived and measured duration.

But the question is deeper.

Is time an independent cosmic road?

Is the universe moving through time like a train moving along tracks?

Is the past physically sitting out there, waiting for telescopes to recover it?

Is redshift truly a doorway into dead history?

Is the expansion of the universe truly the stretching of a fabric, or could it be the field's energy redistribution being read through a time-based model?

These questions cannot be avoided now.

Next part will not simply continue the reinterpretation of objects. It will reinterpret the stage on which those objects were placed.

The cosmic movie reel must be examined.

Because the moment we place the universe inside a movie reel, every observation becomes a frame. Distant galaxies become earlier frames. High redshift becomes older frames. Structure becomes judged by frame number. If something looks mature in an early frame, it becomes anomalous.

But if the universe is not a movie reel, the anomaly changes meaning.

Perhaps JWST has not shown us that the early universe was impossible.

Perhaps it has shown us that "early" was too simple.

Perhaps Hubble tension is not only a disagreement about expansion rate.

Perhaps it is a sign of state-dependent alignment delay.

Perhaps cosmic expansion is not the mechanical stretching of empty space.

Perhaps it is the large-scale observation of energy redistribution through Structured Vacuum Energy.

Perhaps time is not the ruler of the universe.

Perhaps time is the measurable delay between field conditions and resolved consequence.

This is where Alignment Threshold Time becomes essential.

If energy states do not resolve instantly into measurable consequence, then delay appears. If different field regions carry different resistance, strain, density, or saturation, then their alignment delays may differ. What we call time, rate, expansion, and cosmic age may be connected to this deeper process of resolution.

In that sense, time is not denied.

Time is repositioned.

It is not the cosmic road.

It is the field's delay in becoming measurable.

This changes the meaning of the universe completely.

The universe may be fundamentally still in its totality, yet active in its consequences. It may not be rushing through time. It may be resolving field conditions through structured energy behavior. The mind, receiving these consequences in sequence, arranges them into past, present, and future.

That arrangement is necessary for human life.

But it may not be the deepest truth of the cosmos.

This is why next part is called **The End of the Cosmic Movie Reel**.

It does not mean the end of science.

It means the end of treating the universe as a film strip of dead frames. It means the end of assuming that every distant signal is simply an old photograph. It means the end of forcing cosmic depth into developmental childhood. It means the end of allowing time to dominate interpretation without being questioned.

The next chapters must move carefully.

First, we will examine Hubble tension through Alignment Threshold Time. Then we will ask whether time is truly a cosmic road. Then we will explore the still universe and the moving mind. Finally, we will return to the central claim of this book: cosmic anomalies may end when interpretation changes.

The universe did not become strange.

Our reading became exposed.

JWST did not break the cosmos.

It broke the comfort of the movie reel.

And once the signal stops being treated only as history, the universe begins to appear differently.

Not as a dead archive behind us.

But as a layered, structured, equilibrium-seeking present still revealing itself.

Part III — The End of the Cosmic Movie Reel

"Perhaps we were never looking backward into a dead past. Perhaps we were learning to read the depth of a living field."

Chapter 9 — Hubble Tension and Alignment Threshold Time

Why the Hubble tension may not be a crisis, but evidence of state-dependent resolution delay.

When the Universe Refuses One Rate

A crisis begins when reality refuses the simplicity of our measurement. For many years, modern cosmology has tried to describe the expansion of the universe through a clean, elegant, measurable rate. The idea itself is powerful. If the universe is expanding, then we should be able to measure how fast it is expanding. If we know the rate, we can estimate the age, scale, history, and future of the cosmos. One number begins to carry the weight of an entire cosmic story.

That number became deeply important.

It was not merely a measurement. It became part of the identity of the universe. It helped modern cosmology arrange the grand timeline: beginning, expansion, cooling, structure formation, galaxies, stars, planets, observers. The expansion rate became one of the main pillars holding the cosmic movie reel together.

But then the universe did something uncomfortable.

It refused one clean rate.

Different methods began giving different answers. Measurements based on the local universe did not perfectly match values inferred from the early-universe-based model. The difference was not easily disappearing as a simple error. It remained. It became sharper. It became known as the Hubble tension.

The word "tension" is interesting.

It does not say collapse. It does not say failure. It does not say destruction. It says tension. Something is stretched between two readings. Something does not align. Something in the interpretation is under pressure.

Mainstream cosmology naturally treats this as a serious problem. If the universe has one expansion history, and if our model is reading that history correctly, then different methods should eventually converge. Local measurements and early-universe-based calculations should agree within acceptable limits. If they do not, either the measurements are wrong, the assumptions are incomplete, or the model needs modification.

This is why Hubble tension has become so important.

It is not just a numerical disagreement. It is a philosophical disturbance. It tells us that the universe may not be submitting to the neat expansion narrative we placed upon it.

But Universal Energy Dynamics reads this disturbance differently.

Perhaps the universe is not contradicting itself.

Perhaps we are measuring different field states as if they should produce one identical rate.

This is the key shift.

In the standard view, expansion is often imagined as a broad stretching of space, a cosmic scale increasing with time. But in UED, what we call expansion may not be the mechanical stretching of empty spacetime. It may be the large-scale observational expression of energy redistribution, vacuum gradient flow, field resistance, and alignment delay within Structured Vacuum Energy.

If the universe is a structured field, then different regions may not resolve in the same way. A locally stabilized field environment may produce one measurement. A high-redshift, high-resistance, strained field layer may produce another. The difference may not be random. It may be state-dependent.

This is where Alignment Threshold Time becomes essential.

Time, under UED, is not treated as an independent cosmic road on which the universe travels. It is the measurable delay required for energy states to rearrange, align, and become observable as consequence. When the field condition changes, the delay changes. When resistance differs, resolution differs. When ripple tension varies, the apparent rate of cosmic behavior may also vary.

Then Hubble tension begins to look different.

It may not be a crisis of the universe.

It may be evidence of state-dependent resolution delay.

The local universe and the deep field may not be offering two contradictory expansion rates. They may be offering two different readings of alignment through different field conditions. One belongs to our relatively stabilized observational environment. The other belongs to highly strained, high-resistance layers interpreted through the early-universe model.

If we force both readings into one uniform expansion story, we create tension.

But if we read them as field-state differences, the tension becomes information.

This does not mean we should casually dismiss existing measurements. It means we must understand what is actually being measured. Are we measuring pure expansion? Or are we measuring expansion mixed with field resistance, redshift strain, signal delay, and alignment threshold effects? Are we seeing one universal speed? Or are we seeing different resolution behaviors inside a layered field reality?

The question matters because a measurement is never independent of interpretation.

The telescope observes.

The instrument records.

The model translates.

The crisis appears in the translation.

If the universe is treated as an expanding fabric, Hubble tension becomes a frightening crack. But if the universe is treated as a structured field seeking equilibrium through different states of resistance and alignment, Hubble tension becomes a clue. It tells us that cosmic rate may not be independent of field condition.

This chapter begins there.

The universe may not be refusing one rate because it is confused.

It may be refusing one rate because one rate was too simple.

A field reality cannot always be reduced to a smooth cosmic number. Different states resolve differently. Different tensions delay differently. Different regions carry different resistance. If we ignore this and demand a single clean expansion story, the universe will naturally appear inconsistent.

But perhaps the inconsistency is ours.

The universe is not expanding wrongly.

We may be reading alignment delay as expansion rate.

And that is why Hubble tension may not be the collapse of cosmology.

It may be the beginning of a more mature interpretation of cosmic time.

9.1 — What the Hubble Tension Actually Reveals

The Hubble tension is often presented as a technical disagreement.

At first glance, it appears to be a problem of numbers. One method gives one value for the expansion rate of the universe. Another method gives a different value. Scientists check the instruments, refine the calculations, improve the data, and hope that the disagreement will eventually shrink.

But the Hubble tension is not merely a numerical inconvenience.

It is a conceptual disturbance.

It reveals that the universe may not be as easily reducible to one smooth expansion history as the standard model expects. It exposes the hidden confidence that cosmic behavior can be measured through one uniform framework and arranged into one clean timeline. It asks whether different observational methods are truly measuring the same thing, or whether they are measuring different layers of cosmic behavior and forcing them into one category.

In simple terms, the Hubble tension is the disagreement between two major ways of estimating how fast the universe is expanding.

One approach looks at the local universe. It studies relatively nearby cosmic objects, distance ladders, stars, supernovae, and galaxies to estimate how quickly space appears to be expanding around us. This is a more direct, present-near measurement.

Another approach begins with the early-universe model. It uses ancient background signals, theoretical assumptions, and the standard cosmological framework to infer what the expansion rate should be today. This is a more model-dependent calculation, built from the cosmic timeline.

Ideally, both should agree.

If the model is complete and the measurements are clean, the local reading and the early-universe-based reading should converge. They should tell one consistent story. The universe should not give one answer nearby and another answer when inferred from deeper cosmic signals.

But they do not agree perfectly.

That disagreement is the Hubble tension.

Mainstream cosmology treats this seriously because the expansion rate is not a small decorative number. It is connected to the age of the universe, its scale, its structure, its composition, its evolution, and its future. If the expansion rate is uncertain, then the entire cosmic story carries uncertainty. The timeline itself becomes less stable.

This is why the tension matters.

It is not just about how fast galaxies appear to be moving away from one another. It is about whether the model that converts observation into cosmic history is fully reliable. It is about whether the universe can be read through one smooth expansion narrative, or whether something deeper is being missed.

At first, scientists naturally hoped the tension would be resolved through better measurement. Perhaps one method had hidden systematic errors. Perhaps distance measurements needed refinement. Perhaps the early-universe calculation carried some overlooked assumption. Perhaps improved instruments would bring agreement.

That is a reasonable scientific response.

But when a tension persists, it deserves deeper attention.

A persistent tension is not noise. It is a signal.

This does not mean every unresolved disagreement immediately proves a new theory. That would be careless. But it does mean that the disagreement should not be dismissed as a small technical irritation. It may be pointing toward a deeper limitation in the framework itself.

Universal Energy Dynamics reads the Hubble tension from this angle.

Under UED, the disagreement may not mean the universe is confused. It may mean we are reading different field conditions as if they should produce one identical expansion rate.

This is the key point.

The local universe and the deep-field universe may not be equivalent measurement environments. Our local region may represent a relatively stabilized field condition. High-redshift signals and early-universe-based readings may involve strained, resistant, compact, or deeply layered field conditions. If the field states differ, then the way signals resolve through them may also differ.

The standard model assumes that these differences can be smoothed into a single expansion history.

UED asks whether that smoothing is the problem.

If redshift carries field strain, if distance includes field depth, if gravity is energy-gradient equilibrium, and if time is alignment delay, then the Hubble tension becomes more than a discrepancy in expansion rate. It becomes evidence that cosmic measurement is state-dependent.

Different field states may produce different resolution delays.

Different resolution delays may produce different apparent rates.

Different apparent rates may look like a crisis only when forced into one uniform model.

In that sense, the Hubble tension may be revealing the limitation of the cosmic movie reel. The standard model wants the universe to unfold through one continuous sequence of expansion. But a structured field reality may not behave like a film running at one speed. It may resolve differently across different layers, resistances, densities, and alignment conditions.

The tension, then, is not necessarily in the universe.

The tension is in the expectation of uniformity.

This is why the Hubble tension is so valuable. It exposes the difference between measurement and interpretation. The measurement says there is disagreement. The interpretation decides whether that disagreement is an error, a crisis, a patchable parameter, or a clue to a deeper mechanism.

UED chooses the fourth option.

It treats the Hubble tension as a clue.

It asks whether cosmic expansion is truly the stretching of empty spacetime, or whether it is the large-scale appearance of energy redistribution through Structured Vacuum Energy. It asks whether redshift-based readings include field resistance. It asks whether local and deep measurements are affected by different alignment thresholds. It asks whether what we call expansion rate is partly the visible expression of resolution delay.

This does not make the Hubble tension disappear by magic.

It changes what the tension means.

Instead of asking only, "Which measurement is correct?" we must also ask, "What is each measurement actually reading?"

The local measurement may be reading the rate of resolution in our nearby field condition.

The early-universe-based reading may be reading a model-translated value from a different field-state layer.

The difference between them may be the difference between field states, not simply the difference between right and wrong numbers.

This is the deeper revelation.

The universe may not be offering one simple expansion reading because the universe may not be a smooth expanding fabric. It may be a structured field, resolving energy differences through state-dependent delays. If so, then Hubble tension is not an embarrassment. It is a window.

It shows that rate is not independent of condition.

It shows that measurement is not free from field state.

It shows that time, redshift, distance, and expansion may be connected through alignment behavior rather than through chronology alone.

That is why the Hubble tension matters for this book.

It is not merely another anomaly in cosmology.

It is the point where the old cosmic timeline begins to tremble at the level of measurement itself. It tells us that the universe may not be expanding according to one clean story. Or more precisely, it tells us that what we call expansion may be a simplified reading of a deeper field-resolution process.

The Hubble tension does not merely reveal a disagreement between methods.

It reveals the hidden assumption that both methods should speak the same language.

Perhaps they do not.

Perhaps one is reading a stabilized local field.

Perhaps the other is reading a strained deep-field condition.

And perhaps the difference between them is not a crisis to be feared, but a signal to be understood.

9.2 — The Assumption of Uniform Cosmic Expansion

The Hubble tension becomes a crisis only because we expect the universe to give one clean expansion story.

This expectation is hidden so deeply inside modern cosmology that it often appears natural. The universe is assumed to be expanding. This expansion is assumed to follow a broad, smooth, measurable history. Different observations may come from different directions, distances, and methods, but eventually they should fit into one consistent framework. If they do not, something must be wrong.

This is the assumption of uniform cosmic expansion.

It does not necessarily mean every region of the universe is expected to be perfectly identical in every small detail. Modern cosmology does account for local variations, density differences, gravitational effects, and structure. But at the largest scale, the universe is treated as sufficiently uniform and smooth. The expansion is expected to be described through a common framework, one that allows local measurements and deep-field calculations to belong to the same grand cosmic curve.

This assumption gives the standard model its elegance.

It allows the universe to be imagined as one expanding whole. It allows redshift, distance, age, background radiation, galaxy distribution, and cosmic history to be placed into one coherent mathematical story. It allows the universe to become readable as a large-scale structure moving through time.

But every elegance has a cost.

The cost of this elegance is that the universe must behave smoothly enough to protect the model.

When local measurements and early-universe-based calculations disagree, the tension becomes uncomfortable because the disagreement threatens this smoothness. It suggests that the universe may not be giving the same expansion answer from every interpretive direction. It suggests that the rate may not be as universal, clean, or model-independent as expected.

Mainstream cosmology naturally tries to resolve this by searching for errors, adjustments, or hidden parameters. Perhaps the local measurements need correction. Perhaps the early-universe model needs modification. Perhaps dark energy behaves differently. Perhaps there is some new physics hidden inside the gap.

These are valid possibilities.

But Universal Energy Dynamics asks a deeper question.

What if the expectation of one smooth expansion reading is itself incomplete?

What if the universe is not an empty fabric expanding uniformly, but a structured field resolving through different states?

If the vacuum is Structured Vacuum Energy, then cosmic regions are not merely points inside an expanding container. They are field conditions. Some regions may be locally stabilized. Some may be highly strained. Some may be compact. Some may be resistant. Some may be near saturation. Some may carry different ripple tension and alignment delay.

In such a universe, why should every measurement reduce perfectly to one universal expansion rate?

This is where the assumption begins to weaken.

A layered field reality may not behave like a smooth balloon. The balloon metaphor is useful because it gives the mind a picture: space stretches, galaxies move apart, and the expansion rate can be measured. But the universe may not be a balloon. It may not be a passive surface expanding equally in every meaningful sense. It may be a field of structured energy, where apparent expansion reflects redistribution, gradient flow, resistance, and alignment behavior.

If different regions carry different field states, then signals emerging from those regions may resolve differently. Their redshift may include different levels of strain. Their delay may include different resistance. Their apparent distance may include field depth. Their measured expansion may include alignment threshold effects.

Then the Hubble tension is not surprising.

It is expected.

The crisis begins when a layered field reality is forced into a single expansion narrative.

A simple analogy may help. Imagine measuring the flow of water in a river. If the river is calm, uniform, and open, one rate may describe it reasonably well. But if the river contains rocks, bends, narrow channels, whirlpools, and depth variations, different points may produce different flow readings. A single number may still be useful as an average, but it will not capture the full condition of the river.

If someone insists that every measurement must match one smooth rate, the river will appear contradictory.

But the contradiction is not in the river.

It is in the expectation.

Similarly, the cosmos may not be a smooth flow of expansion. It may contain field-state differences across scales. Local measurements may read one condition. Deep redshift-based readings may read another. The disagreement may not simply mean one method is wrong. It may mean the universe is not being measured as a uniform expansion system, but as a structured field with different resolution states.

This also changes how we understand the word "rate."

A rate assumes a process unfolding against time. But if time itself is not an independent cosmic road, and instead represents alignment delay, then the measured rate may depend on how quickly field conditions resolve. Different field states may produce different resolution speeds. A stable local field may align differently from a strained deep field. A low-resistance region may resolve differently from a high-resistance region.

Therefore, what appears as expansion rate may partly be resolution behavior.

This is the UED shift.

The standard model asks: what is the true expansion rate of the universe?

UED asks: what field-state resolution is being measured as expansion?

The standard model asks: why do local and early-universe measurements disagree?

UED asks: why should different field states produce identical alignment readings?

The standard model sees the disagreement as a problem for uniform expansion.

UED sees it as evidence against over-simplified uniformity.

This does not mean the universe is chaotic. It means the universe is structured. A structured system can contain different states while still belonging to one whole. The human body has one identity, but temperature, pressure, oxygen levels, electrical activity, and cellular behavior vary across regions. A planet has one identity, but its atmosphere, oceans, core, magnetic field, and surface conditions are not uniform. A galaxy has one identity, but its arms, center, halo, and star-forming regions behave differently.

Why should the universe be forced into smoother simplicity than the systems inside it?

Uniformity is useful as a broad approximation. But approximation becomes dangerous when it becomes unquestioned truth. If the universe is fundamentally a field reality, then large-scale unity does not require identical state behavior everywhere. Unity can exist through relation, not sameness. The field may be continuous, but its conditions may vary. Its resolution delays may differ. Its observable rates may therefore differ.

Hubble tension may be the signal of this deeper fact.

It may be telling us that the universe is not expanding like a simple uniform fabric. It may be telling us that cosmic measurement depends on the condition of the field being measured. It may be telling us that redshift, distance, and rate are not independent of strain, resistance, and alignment delay.

This is why UED does not panic before Hubble tension.

It listens to it.

The tension may be saying that the universe cannot be reduced to one smooth expansion story without losing field reality. It may be saying that the local universe and the deep-field universe are not merely different positions on one timeline, but different resolution conditions within one structured present.

The old model expects uniformity.

The universe may be offering state-dependence.

And perhaps the true crisis is not that the universe refuses one rate.

The true crisis is that we expected one rate to carry the whole universe.

9.3 — Time as Alignment Threshold, Not Cosmic Road

Time is one of the most useful ideas in human experience.

It helps us live. It helps us remember. It helps us plan. It helps us compare one event with another. It allows us to say before, after, early, late, fast, slow, past, present, and future. Without time, ordinary life would lose its arrangement.

But usefulness does not automatically make time fundamental.

This is where cosmology must become careful.

In ordinary thinking, time is imagined as a road. Events move along it. The universe travels through it. The past lies behind, the future lies ahead, and the present is the moving point between the two. This image is so natural to the human mind that we rarely question it. We say the universe began in time, evolved through time, expanded over time, and will continue into the future.

Time becomes the invisible highway of existence.

But Universal Energy Dynamics does not begin this way.

UED does not treat time as an independent cosmic road on which reality travels. It treats time as **Alignment Threshold Time**, or ATT: the measurable delay required for energy states to rearrange, interact, align, and become observable as consequence.

This is a very different understanding.

In this view, time is not the master of events. Time is the measurable delay created by events as they resolve. It is not something outside the universe carrying change. It is the interval experienced or measured when energy conditions move from one state toward another.

Where there is no change, no interaction, no disturbance, no resistance, and no resolution, time has no meaningful role.

Time appears because something is happening.

More precisely, time appears because something takes duration to resolve.

A simple example can help.

If we heat water, it does not immediately become steam. Energy enters the system. Molecules respond. Temperature rises. Internal conditions change. At a certain threshold, boiling begins. The duration between heating and boiling is not a mysterious road through which water travels. It is the measurable delay required for the system to reach a new state.

That delay is real.

But it is not independent of the process.

Similarly, if a piece of metal is heated, it expands according to its own material condition. If wood is heated, it responds differently. If water is heated, it responds differently again. Each system has its own response, resistance, threshold, and delay. We measure time across these processes, but the delay depends on the state of the system.

This is the basic spirit of Alignment Threshold Time.

Different systems do not resolve identically because their conditions are different. Their internal resistance is different. Their density is different. Their structure is different. Their thresholds are different. Therefore, their measurable delays are different.

Now apply this to the cosmos.

If the universe is treated as empty space filled with matter, then time is easily imagined as a universal background. It flows everywhere, while objects and signals move through it. But if the universe is a Structured Vacuum Energy field, then every observed process involves field condition. Energy must resolve through resistance, tension, density, gradient, and alignment. Time becomes the delay of this resolution.

This means cosmic time cannot be separated from field state.

A locally stabilized field condition may resolve in one way. A high-strain field region may resolve differently. A compact high-resistance layer may produce a different delay. A saturated field zone may resist coherence in another way. Signals emerging from or passing through these conditions may not carry identical alignment behavior.

This is where ATT becomes essential for understanding Hubble tension.

The Hubble tension appears because different methods give different expansion readings. The standard interpretation treats this as a problem because it expects one smooth cosmic expansion history. But if time itself is alignment delay, then different field states can naturally produce different apparent rates. The measured rate may not be pure expansion. It may be expansion-like interpretation mixed with state-dependent delay.

In simple words, different regions may not be "expanding differently" in a confused way.

They may be resolving differently.

This changes the question.

Instead of asking, "Why does the universe have two expansion rates?" we ask, "Are we measuring different alignment thresholds and mistaking them for one universal expansion rate?"

That question is more fundamental.

It moves the discussion from cosmic speed to cosmic resolution.

Under UED, ATT is the time required for an energy condition to become measurable after interacting with the field. If the field is resistant, the delay changes. If the field is strained, the signal changes. If the field is compact, the resolution changes. If the field is saturated, coherence may fail entirely.

Therefore, what we call cosmic rate may depend on how field states resolve into observation.

This does not deny clocks. It does not deny measurement. It does not deny duration. It simply denies that time is an independent cosmic ruler standing above reality. Clocks measure regular processes. They do not prove that time exists as a physical road outside process. A clock is useful because its internal process is stable enough to compare with other processes.

But comparison is not fundamentality.

A clock measures change by using change.

Similarly, cosmology measures cosmic behavior through signals, redshift, distance ladders, background patterns, and models. These measurements are valuable. But if the signals themselves are shaped by field state and alignment delay, then the measured time-related quantities must be interpreted carefully.

This is why ATT matters.

It gives us a way to understand time without turning it into an invisible god.

Time is not denied.

Time is repositioned.

It becomes the delay of field resolution.

It becomes the measurable interval between disturbance and alignment.

It becomes the consequence of energy states trying to settle into observable form.

Once this is understood, Hubble tension looks less like a cosmic contradiction and more like a measurement of different resolution conditions. The local universe may reflect one alignment threshold. Deep-field or early-universe-inferred measurements may reflect another. The difference between them may reveal the structure of the field, not the failure of the universe.

The old model says: the universe should have one expansion rate.

UED says: different field states may carry different alignment delays.

The old model says: the disagreement is a crisis.

UED says: the disagreement may be a signal.

The old model asks time to behave like a road.

UED reads time as the footprint of resolution.

This is the key insight.

Time is not the ruler of cosmic change.

Time is the delay created when cosmic change becomes measurable.

And when the delay depends on field condition, the universe will not always submit to one clean rate.

9.4 — State-Dependent Resolution Delay

If time is alignment delay, then delay cannot be identical everywhere.

This is the key to understanding Hubble tension through Universal Energy Dynamics.

In the usual interpretation, the universe is expected to follow one broad expansion history. Different measurements may come from different regions, different instruments, and different methods, but ultimately they are expected to converge into one clean rate. If they do not, the disagreement becomes a crisis.

But UED begins with a different assumption.

The universe is not an empty fabric expanding uniformly. It is a structured field reality. Different regions of this field may carry different states of resistance, strain, density, compactness, saturation, and alignment. If the field states are different, then the delay required for energy conditions to resolve into measurable consequence will also be different.

This is state-dependent resolution delay.

It means that what we measure as time, rate, redshift, distance, or expansion may depend on the condition of the field through which the signal or process is resolving. A stable field state and a strained field state cannot be expected to behave identically. A low-resistance region and a high-resistance region cannot be expected to produce the same delay. A locally settled field and a deep high-tension field layer cannot be treated as if they are merely two points on the same smooth clock.

The field state matters.

This is easy to understand in ordinary systems.

Different materials respond differently to heat. Metal conducts heat quickly. Wood responds differently. Water absorbs heat and changes state at a threshold. Air expands differently. Each system has its own internal structure, resistance, and response delay. If we apply the same energy input to different materials, they do not resolve in the same way.

The difference is not an error.

It is state-dependence.

Similarly, if we disturb different systems, they settle at different speeds. A calm pond settles differently from a stormy sea. A flexible bridge responds differently from a rigid wall. A dry soil absorbs water differently from saturated soil. The input may be similar, but the response depends on condition.

Why should the cosmos be exempt from this principle?

If the universe is Structured Vacuum Energy, then field conditions cannot be ignored. Some regions may be relatively stable from our observational frame. Some may be deeply strained. Some may be compact and resistant. Some may exist near saturation boundaries. Some may produce signals that carry heavy ripple tension. Their alignment delays will not be identical.

Now apply this to the Hubble tension.

Local measurements come from our relatively nearby cosmic environment. They are based on signals, objects, and calibrations within a field condition closer to our observational state. Early-universe-based calculations, or deep-field interpretations, are built through signals that have emerged from or passed through very different field conditions. These signals may carry stronger strain, resistance, and alignment delay.

If both readings are forced into one universal expansion rate, disagreement appears.

But if they are understood as readings of different field-state delays, the disagreement becomes meaningful.

The local reading may represent one resolution condition.

The deep-field reading may represent another.

The difference between them may not be a flaw in the universe. It may be the observational imprint of different alignment thresholds.

This is why Hubble tension may be more than a crisis. It may be evidence that cosmic measurement is state-dependent.

The standard model treats the two readings as if they should describe one smooth expansion history. UED asks whether they are actually describing different resolution states within the same structured field. The model expects uniformity; the field may be revealing layered behavior.

This also changes how we understand expansion.

If expansion is treated as the stretching of empty spacetime, then the disagreement feels severe. One universe should not give two expansion answers. But if apparent expansion is partly the large-scale expression of energy redistribution, vacuum gradient flow, and alignment delay, then different readings are not shocking. They may reveal that the field does not resolve uniformly across all conditions.

In that case, expansion rate is not purely a cosmic speed.

It is also a measurement of field response.

This is where ATT becomes central. Alignment Threshold Time tells us that measurable duration appears when energy states require delay to resolve. If the delay changes with field condition, then the measured rate changes with field condition. The same universe may appear to expand differently depending on which field-state layer is being read.

This does not mean everything becomes arbitrary.

State-dependence is not chaos. It means there is structure beneath the variation. Differences are not random; they arise from condition. If we understand the condition, we can understand the difference. If we ignore the condition, the difference becomes a crisis.

Hubble tension may therefore be the universe asking us to include field condition in cosmological measurement.

It may be saying: do not read all redshift as simple distance.

Do not read all delay as simple history.

Do not read all rate as uniform expansion.

Do not read all disagreement as error.

Read the state.

This is a more mature approach because it does not panic before disagreement. It asks what the disagreement is revealing. It treats tension as information. It understands that when two measurements do not match, the first task is not to force one into the other, but to ask whether both are measuring the same underlying condition.

Perhaps they are not.

The local universe and deep-field universe may not be different chapters of a smooth movie reel alone. They may be different field states with different resolution delays. The local field may be closer to equilibrium from our observational position. The deep field may carry higher resistance, stronger strain, and longer alignment delay. When these are interpreted through one expansion model, the result appears as Hubble tension.

But through UED, it becomes state-dependent resolution.

This also prepares a deeper understanding of time itself. If time were an independent cosmic road, then every process would be placed on that road and compared through one common framework. But if time is alignment delay, then the condition of the field becomes inseparable from the measurement of duration. There is no pure time floating outside field behavior. There is only resolution measured as time.

This is why Hubble tension is so important for the end of the cosmic movie reel.

A movie reel runs frame after frame at one sequence. If the universe is imagined like that, every measurement must fit into the same temporal unfolding. But a field does not behave like a film strip. It resolves according to condition. Different regions may carry different delays. Different signals may carry different strain. Different observations may reveal different thresholds.

The movie reel expects one clock.

The field reveals many resolutions.

This is not contradiction.

It is depth.

State-dependent resolution delay therefore converts Hubble tension from a frightening anomaly into a meaningful clue. It tells us that the universe may not be governed by one smooth expansion story. It may be a layered field where apparent rates emerge from the interaction between energy states, resistance, and observational resolution.

The disagreement is not the universe disagreeing with itself.

It is the universe refusing to be reduced to one oversimplified rate.

9.5 — Redshift, ATT, and the Misreading of Expansion

Redshift sits at the center of the expansion story.

A signal arrives stretched toward red. The standard interpretation reads this stretching as evidence of cosmic expansion. The farther the object, the greater the redshift. The greater the redshift, the deeper the lookback time. The deeper the lookback time, the more confidently the object is placed inside the early universe.

This chain has become the backbone of modern cosmological interpretation.

But if redshift carries field strain, and if time is Alignment Threshold Time, then this chain can no longer remain untouched.

Redshift may not be only the mark of expanding space.

It may also be the mark of delayed field resolution.

This is where the misreading of expansion begins.

In the standard model, expansion is treated as a large-scale stretching of space. Light travels through that expanding space, and its wavelength stretches along the way. The resulting redshift becomes a sign that the universe has expanded during the signal's journey. This creates a clean interpretation: redshift is expansion history written into light.

But Universal Energy Dynamics asks whether this is complete.

If the vacuum is not empty, but Structured Vacuum Energy, then a signal does not travel through passive nothingness. It resolves through field condition. It passes through resistance. It carries ripple tension. It is affected by alignment delay. Its wavelength state may reflect not only spatial expansion, but also the difficulty of resolving through different field states.

In this case, redshift becomes more than an expansion marker.

It becomes a field-resolution marker.

This matters because expansion rate is inferred through redshift, distance, and time. If redshift itself contains field strain and ATT effects, then the expansion rate we calculate may not be pure expansion. It may be expansion mixed with alignment delay.

The number may be real.

But its meaning may be incomplete.

This is exactly where Hubble tension becomes meaningful under UED. Local measurements and deep-field or early-universe-inferred measurements may disagree because they do not read the same field-resolution condition. A local signal may pass through a relatively stabilized field environment. A deep redshifted signal may emerge from or pass through high-resistance, high-strain field layers. The redshift in each case may carry different alignment-delay components.

If we interpret both through one uniform expansion model, tension appears.

But if we interpret them through field-state resolution, the disagreement becomes expected.

The problem is not that the universe is expanding at two contradictory rates.

The problem may be that we are reading different alignment delays as if they were one universal expansion rate.

This is a very important correction.

Expansion, in UED, is not denied as an observed large-scale behavior. What is questioned is the physical meaning we attach to it. What we call expansion may be the observable expression of energy redistribution across Structured Vacuum Energy. It may be vacuum gradient flow. It may be field relaxation across scale. It may be the movement of measurable consequences through different alignment thresholds.

In other words, expansion may be less like a fabric stretching and more like a field resolving.

If so, redshift is not merely the stretching mark of space.

It is the signal's trace of resolution through the field.

This also changes how we interpret high-redshift regions. In the old reading, high redshift means strong lookback, strong distance, and early stage. In UED, high redshift may also mean stronger field resistance, stronger ripple tension, and greater alignment delay. A highly redshifted structure may not only be far and delayed; it may be field-deep and strained.

Then the Hubble tension begins to speak more clearly.

The local universe may show one apparent expansion rate because the local field condition resolves in one way. Deep-field readings may imply another because the signals involved carry different field-strain and ATT effects. The disagreement may not be a flaw in observation. It may be a clue that cosmic rate is state-dependent.

The standard interpretation asks:

Why do the expansion rates disagree?

UED asks:

What field states are hidden inside those rates?

The standard interpretation asks:

Which measurement is wrong?

UED asks:

Which delay is being mistaken for expansion?

The standard interpretation asks:

Do we need a new parameter?

UED asks:

Do we need a new meaning of time, redshift, and field resistance?

This is the deeper path.

When redshift is read only through expansion, it becomes a cosmic clock. When redshift is read through ATT, it becomes a sign of how long a field condition takes to resolve into observable consequence. That difference is not cosmetic. It changes the entire interpretation of cosmic measurement.

A clock tells us sequence.

A resolution delay tells us condition.

The Hubble tension may therefore be warning us that the universe is being timed incorrectly. Not in the simple sense that our clocks are broken, but in the deeper sense that the model treats time as a universal road instead of a field-dependent delay. If time is ATT, then every cosmological rate must be examined through the field condition from which it is derived.

Redshift cannot be separated from ATT.

Distance cannot be separated from field depth.

Expansion cannot be separated from energy redistribution.

Rate cannot be separated from resistance.

This integrated reading is what mainstream cosmology often misses because it separates categories too cleanly. It treats redshift as one thing, distance as another, time as another, expansion as another, and structure as another. Then it tries to connect them through equations. UED begins from a more unified foundation: all these are consequences of structured energy resolving through field conditions.

So, when the readings disagree, UED does not immediately panic.

It asks what kind of field behavior the disagreement is revealing.

Hubble tension may be the result of reading redshift as simple expansion when it is partly field strain. It may be the result of treating lookback time as cosmic history when it is partly alignment delay. It may be the result of assuming distance is only separation when it may also be field depth.

The universe may not be giving two different answers.

We may be asking one question where two layers are involved.

One layer is spatial-observational.

The other is field-resolution.

When both are collapsed into a single expansion rate, the mismatch becomes visible.

This is why Hubble tension is not merely a problem to be patched. It is a clue to the wrong compression of meaning. We compressed redshift into distance. We compressed distance into time. We compressed time into history. We compressed history into expansion rate. Then reality resisted the compression.

Under UED, the resistance itself is informative.

It tells us that cosmic observation is layered.

It tells us that signals carry field state.

It tells us that time is delay, not road.

It tells us that expansion may be interpretation, not pure mechanism.

This does not mean every existing cosmological calculation must be discarded. It means the conceptual foundation must be widened. Calculations may still be useful within defined ranges, but their physical interpretation must include field condition. Without ATT, redshift may be overburdened. Without SVE, expansion may be oversimplified. Without field resistance, Hubble tension may be misunderstood.

The deeper issue is not the number alone.

The deeper issue is what the number is assumed to represent.

If redshift is only expansion, Hubble tension is a crisis.

If redshift includes field strain and ATT, Hubble tension becomes a diagnostic signal.

It tells us that different cosmic measurements are not merely disagreeing; they are revealing different layers of resolution within the structured field.

The universe may not be expanding wrongly.

It may be resolving differently.

And the redshift we trusted as a simple cosmic clock may actually be the first visible sign that time, distance, and expansion are all field-dependent consequences.

9.6 — Hubble Tension as Confirmation, Not Collapse

A crisis can look very different when the frame changes.

Inside the standard cosmological frame, the Hubble tension appears alarming. It suggests that the universe is not giving one consistent expansion answer. Local measurements point in one direction. Early-universe-based calculations point in another. The difference is not large in ordinary imagination, but in cosmology, even a small persistent difference can shake a large structure.

Naturally, the word crisis appears.

But crisis is not always collapse.

Sometimes crisis is confirmation that an older frame has become too narrow.

This is how Universal Energy Dynamics reads the Hubble tension. It does not treat the disagreement as proof that the universe is broken. It treats the disagreement as evidence that cosmic measurement may be state-dependent. In other words, the universe may not be offering two contradictory expansion rates. It may be showing that different field conditions produce different resolution delays, and those delays are being misread as one uniform expansion rate.

This is not a minor correction.

It changes the entire emotional meaning of the tension.

The standard model expects one smooth expansion narrative. It expects the local universe and the deep-inferred universe to align under one broad mathematical story. When they do not, the model feels stressed. But UED does not begin with smooth empty expansion. It begins with Structured Vacuum Energy, field resistance, ripple tension, energy redistribution, and Alignment Threshold Time.

Inside this framework, difference is not automatically failure.

Difference can be information.

If a local field condition is relatively stabilized, the signals and measurements derived from it may resolve through one alignment behavior. If high-redshift or deep-field signals emerge through strained, compact, resistant, or saturated conditions, their resolution delay may differ. If both are converted into a single expansion framework, tension appears. But the tension is not necessarily a defect. It may be the field revealing its layered nature.

This is why Hubble tension may be confirmation.

Not simplistic proof.

Not final victory.

Not a cheap declaration that one framework has defeated another.

But confirmation in the sense that a persistent discrepancy points toward exactly the kind of layered, state-dependent reality that UED expects.

A uniform-fabric universe should ideally give one clean expansion reading.

A structured-field universe may not.

A cosmic movie reel should run through one orderly sequence.

A field reality may resolve according to condition.

This is the difference.

If the universe is treated as a film strip moving through time, then all frames must belong to the same timeline. The early frame, middle frame, and present frame must fit into one clean progression. Hubble tension disturbs this because the early-inferred and present-measured values do not settle neatly into one frame sequence.

But if the universe is a field of layered resolution, then different readings may reflect different alignment states. The disagreement does not destroy the universe. It exposes the weakness of the movie reel.

This is why the Hubble tension belongs at the beginning of this part. It is not only a problem of expansion. It is a problem of time. It asks whether time is truly a cosmic road, or whether what we call time is a measured delay of field resolution. If time is ATT, then expansion rate cannot be separated from alignment behavior. And if alignment behavior depends on field state, then rate itself becomes state-dependent.

This makes the tension meaningful.

The universe may be saying: do not reduce me to one clock.

Do not reduce redshift to one timestamp.

Do not reduce distance to one history.

Do not reduce expansion to one smooth stretching.

Read the field.

This does not mean every disagreement in cosmology should be celebrated as confirmation. That would be careless. Measurement errors still matter. Calibration matters. Models must be tested. Instruments must be refined. UED must also face scrutiny. A serious framework cannot simply collect every anomaly and claim success.

But the Hubble tension is special because it sits at the intersection of redshift, distance, time, expansion, and cosmic history. It is not an isolated irregularity. It touches the core assumptions of the cosmic movie reel. It asks whether the universe is being read through a model that expects more uniformity than reality provides.

That is why UED takes it seriously.

Under UED, the disagreement between measurements may be the observational trace of different field-state delays. Local measurements may carry the resolution behavior of our nearer stabilized field environment. Early-universe-inferred values may carry assumptions built on signals from deeper, more strained layers. Their mismatch may reveal the difference between local alignment and deep-field alignment.

The old interpretation calls this tension.

UED calls it state-dependent resolution.

Once this shift is made, the question changes completely.

We stop asking only, "Which value is correct?"

We begin asking, "What condition produced each value?"

We stop asking only, "How fast is the universe expanding?"

We begin asking, "What field behavior are we interpreting as expansion?"

We stop asking only, "Why does the cosmic clock disagree?"

We begin asking, "Is there one cosmic clock at all?"

This is the beginning of a deeper cosmology.

A mature science does not panic when measurements disagree. It asks whether the disagreement is noise, error, limitation, or revelation. In the case of Hubble tension, UED suggests that the disagreement may be revelation. It may reveal that cosmic rates are not independent of field conditions. It may reveal that redshift carries strain. It may reveal that time is delay. It may reveal that expansion is not merely the stretching of emptiness but the appearance of energy redistribution through Structured Vacuum Energy.

This is why Hubble tension does not weaken the UED lens.

It strengthens the need for it.

The tension tells us that the old frame may be too smooth. It tells us that one universal expansion reading may be too simple. It tells us that the universe may not behave like a single cinematic sequence projected across spacetime. It tells us that different regions and signals may carry different histories of resistance, strain, and alignment.

The cosmos may not be inconsistent.

It may be layered.

The rate may not be wrong.

It may be incomplete.

The crisis may not be in reality.

It may be in the expectation that reality must resolve through one clean number.

This is the strongest philosophical landing of the Hubble tension under UED. The universe is not collapsing into contradiction. The old interpretation is being forced to recognize field complexity. What mainstream cosmology experiences as stress, UED reads as signal. What appears as a crack in the expansion model may be a window into the resolution behavior of the field.

So Hubble tension may not be the universe disagreeing with itself.

It may be the universe refusing to let us mistake alignment delay for universal expansion rate.

And that refusal may be one of the most important clues we have.

Closing Reflection — The Tension Was in the Interpretation

The universe may not be in tension.

Our interpretation may be.

That is the quiet realization this chapter must leave behind. For a long time, we have looked at the Hubble tension as if the universe itself were giving conflicting answers. One measurement says one thing. Another measurement says something slightly different. The expected agreement does not arrive. The smooth story hesitates. The cosmic movie reel begins to stutter.

Naturally, we call it tension.

But perhaps the tension is not in the cosmos.

Perhaps it is in the expectation that the cosmos must behave according to one clean expansion reading.

This expectation is understandable. Science seeks order. It wants convergence. It wants different methods to support the same result. When measurements agree, confidence grows. When they disagree, doubt enters. This is healthy. Without such discipline, science would become storytelling without correction.

But there is another possibility.

Sometimes disagreement is not failure.

Sometimes disagreement is the first honest signal that a deeper layer has been ignored.

The Hubble tension may be such a signal.

If the universe were merely an empty spacetime fabric expanding uniformly, then one universal expansion story would be expected. Local measurements and deep-inferred calculations should eventually align. If they do not, then the model must search for errors, hidden parameters, or new adjustments.

But if the universe is a structured field reality, then the expectation changes.

A structured field does not need to resolve identically everywhere. Different regions may carry different resistance. Different signals may carry different ripple tension. Different field states may have different alignment delays. A locally stabilized field condition may not produce the same observational reading as a deep, strained, high-resistance field layer.

Then the disagreement becomes meaningful.

It becomes the signature of state-dependent resolution.

This is where Alignment Threshold Time changes the entire discussion. Time is no longer treated as an independent road along which the universe travels. Time becomes the measurable delay required for energy states to align, resolve, and appear as observable consequence. If the field state changes, the delay changes. If the delay changes, the apparent rate changes. If the apparent rate changes, the universe may seem to offer different expansion readings.

But the universe is not contradicting itself.

It is revealing that rate depends on condition.

This is the central insight of the chapter.

The Hubble tension may not be a crisis of expansion. It may be a crisis of interpretation. We may be mistaking alignment delay for expansion rate. We may be compressing redshift, distance, time, resistance, and field strain into a single simplified number. We may be asking one question while the universe is answering in layers.

The old model asks, "How fast is the universe expanding?"

UED asks, "What field condition is being measured as expansion?"

That difference matters.

Because once the question changes, the meaning of the tension changes. The tension is no longer a frightening crack in reality. It becomes a diagnostic point. It tells us that the universe may not be readable through one uniform clock. It tells us that cosmic signals may carry the effect of resistance and strain. It tells us that local and deep-field observations may not belong to one simple resolution state.

This does not make the problem disappear.

It makes the problem more intelligent.

A mature framework does not erase difficulty by renaming it. It gives difficulty a deeper place. Hubble tension still requires measurement, refinement, criticism, and testing. But under UED, it is no longer merely an embarrassment to be patched. It becomes a clue that cosmic time and cosmic expansion are not as independent as we imagined.

If redshift carries field strain, and if distance includes field depth, and if time is alignment delay, then expansion cannot be treated as a simple stretching story. What we call expansion may be the large-scale appearance of energy redistribution through Structured Vacuum Energy. It may be the field resolving imbalance across layers. It may be the universe becoming measurable through different thresholds of alignment.

This is why the cosmic movie reel begins to fail here.

A movie reel assumes sequence. One frame follows another. The entire story runs through a continuous order. If we know where a frame belongs, we know its place in the story. But a field reality is not so simple. A field resolves according to condition. It does not need to behave like a film running at one fixed interpretive speed.

The Hubble tension shows this failure of the reel.

It tells us that the universe may not be moving through time like a story projected across space. It may be resolving through field conditions. What appears as different expansion readings may be different alignment states. What appears as disagreement may be layered truth.

This is not the end of cosmology.

It is the beginning of a more careful cosmology.

A cosmology that does not treat time as untouchable.

A cosmology that does not reduce redshift to a clock.

A cosmology that does not reduce distance to history.

A cosmology that does not reduce expansion to fabric-stretching.

A cosmology that reads field state before declaring crisis.

The Hubble tension may therefore be one of the most important signals modern observation has given us. Not because it proves every detail of UED automatically, but because it demands the kind of deeper framework UED offers. It asks for a universe where rate, resistance, delay, and field condition belong together. It asks for a universe where measurement is not separated from the state of reality being measured.

The tension was never asking us to panic.

It was asking us to think deeper.

Perhaps the local universe and the deep universe are not disagreeing. Perhaps they are speaking from different field conditions. Perhaps one reflects a stabilized resolution state while the other carries the imprint of high-resistance alignment delay. Perhaps both are correct within their conditions, and the mistake lies in forcing both into one smooth expansion narrative.

If that is true, then Hubble tension is not the universe failing consistency.

It is the universe refusing oversimplification.

And that refusal is valuable.

Because science progresses when reality refuses our comfort.

The chapter began with the idea that the universe refuses one rate. But now we can say it more clearly: the universe may not be refusing order. It may be refusing our simplified order. It may be showing that order exists at a deeper level, not as uniform expansion, but as state-dependent resolution.

The tension was not in the universe.

The tension was in the interpretation.

And once interpretation changes, the tension becomes a message:

Stop mistaking alignment delay for universal expansion rate.

Stop forcing field reality into a single cosmic clock.

Stop reading layered resolution as contradiction.

The universe may not be expanding wrongly.

We may have been timing it too simply.

Chapter 10 — Time Is Not a Cosmic Road

Time as causal sequencing, alignment delay, and conscious narration—not an independent road on which the universe travels.

The Road We Mistook for Reality

Time is the most familiar road we never see.

We walk through it without touching it. We measure it without holding it. We fear losing it, regret wasting it, celebrate moments inside it, and arrange our entire existence through it. Morning comes, evening comes, years pass, bodies age, memories accumulate, plans move forward, and death waits somewhere ahead.

So naturally, we believe time is real.

Not merely useful.

Real.

And because it feels so real in experience, we place the entire universe inside it.

We imagine the past behind the universe, the future ahead of it, and the present as the moving point where cosmic reality is happening now. We imagine the universe travelling through time as if time were a road, and existence were a vehicle moving along it. The Big Bang becomes the beginning of the road. Cosmic expansion becomes the journey. Galaxies, stars, planets, life, and consciousness become milestones along the way.

This image is powerful because it matches human experience.

We live narratively.

We remember what happened. We act in the present. We prepare for what may happen. Our minds cannot function without sequence. A life without before and after would be impossible to organize. A memory without past would be meaningless. A plan without future would have no direction. Even science depends on sequence: observation, measurement, comparison, prediction, verification.

So time is not useless.

It is one of the greatest organizing tools of conscious experience.

But usefulness is not the same as fundamentality.

This is the mistake.

Because time is necessary for the mind, we assume it must be necessary for the universe in the same way. Because experience unfolds in sequence, we assume reality itself is travelling along a cosmic sequence. Because we remember past events, we imagine the past exists somewhere as a real cosmic archive. Because we anticipate the future, we imagine the future as a region waiting ahead. Because we measure duration, we assume time is an independent container inside which change happens.

But what if time is not the road?

What if time is the arrangement we create after consequences occur?

What if time is not carrying the universe, but appearing because energy states take duration to resolve?

Universal Energy Dynamics begins with this reversal.

Time is not treated as an independent cosmic highway. It is not the invisible river through which the universe flows. It is not the master container of existence. Time is understood as causal sequencing, alignment delay, and conscious narration.

A cause leads to consequence.

An energy state interacts with another state.

A field condition resists, aligns, saturates, or resolves.

A signal takes delay to arrive.

A process takes duration to complete.

A conscious mind receives these consequences and arranges them as past, present, and future.

That arrangement is what we call time.

This does not deny time in ordinary experience. It does not deny clocks. It does not deny aging. It does not deny signal delay. It does not deny history as a meaningful human construction. But it does remove time from the throne. It stops treating time as an independent ruler of the universe and begins treating it as the measurable and conscious expression of change, delay, and consequence.

This distinction matters deeply for cosmology.

If time is a road, then the universe becomes a movie reel. It begins, develops, matures, and moves forward frame by frame. Distant galaxies become earlier frames. High redshift becomes ancient time. JWST becomes a machine looking backward into the cosmic childhood. A mature structure at cosmic depth becomes an anomaly because it appears in the wrong frame.

But if time is alignment delay and conscious narration, the movie reel begins to weaken.

A delayed signal is not automatically a dead historical scene. A redshifted galaxy is not automatically a child of the universe. Cosmic depth is not automatically cosmic infancy. The universe may not be storing its past like old photographs waiting for telescopes to recover them. It may be resolving field conditions into observable consequences, and the mind may be arranging those consequences as history.

This is why time must be examined.

Not because time is meaningless.

But because time may have been given the wrong status.

The universe may not be travelling through time.

Structured energy may be resolving through conditions.

The field may be aligning, resisting, saturating, redistributing, and becoming measurable through consequences. Consciousness then experiences those consequences in sequence and names the sequence time.

The road may not be outside reality.

The road may be inside narration.

And if that is true, then modern cosmology has made a very human mistake. It has projected the mind's need for sequence onto the whole universe. It has turned delay into history, history into cosmic biography, and cosmic biography into physical certainty.

This chapter begins by questioning that projection.

Time is real as experience.

Time is real as measurement.

Time is real as duration between conditions.

But time may not be real as an independent road on which the universe travels.

Perhaps the universe does not move through time.

Perhaps time appears when the universe moves into consequence.

10.1 — Why the Mind Needs Time

The mind needs time because the mind cannot live inside raw reality.

It must arrange.

It must remember.

It must compare.

It must expect.

It must connect one event with another and say, "This happened before that." Without this arrangement, experience would collapse into scattered impressions. A sound, a face, a pain, a thought, a movement, a fear, a decision — all would appear without order. The mind would receive reality, but it would not understand it.

Time gives experience sequence.

And sequence gives experience meaning.

This is why time feels so fundamental to human life. We do not merely use time to read clocks. We use time to build identity. A person says, "I was born here. I studied there. I struggled then. I changed after that. I became this because of those experiences." The self is built through time. Memory becomes the past. Desire becomes the future. Attention becomes the present.

Without time, even the word "I" would become unstable.

Who are we without memory?

What is responsibility without consequence?

What is learning without before and after?

What is regret without the past?

What is hope without the future?

The mind needs time because the mind is not merely observing reality. It is organizing life.

This is why time becomes emotionally powerful. We fear running out of it. We feel guilty for wasting it. We celebrate anniversaries, birthdays, achievements, deadlines, beginnings, endings, and milestones. Human life is not experienced as one continuous undivided field. It is experienced as chapters. Childhood, youth, adulthood, aging. Yesterday, today, tomorrow. Before success, after failure. Before loss, after realization.

Time becomes the grammar of human existence.

Science also depends on this grammar.

A scientist observes an event, records it, compares it with another event, repeats the process, measures the interval, and builds a law. Cause must precede consequence. A signal must be emitted before it is received. A reaction must begin before it completes. A star must form before it collapses. A measurement must be taken before a conclusion is drawn.

Without sequence, science cannot operate.

So we must be very careful here.

To question time does not mean to insult time. It does not mean time is useless, false, or meaningless. It does not mean clocks are illusions in a silly sense. It does not mean history has no value. It does not mean human experience should ignore duration. Such shallow rejection would create confusion, not clarity.

Time is real in experience.

Time is real in measurement.

Time is real as duration between conditions.

But the question is whether time is fundamental in the way we imagine.

The mind needs time, but does the universe need time in the same way?

This is the deeper question.

The mind arranges consequences into a story because the mind cannot process all reality at once. It receives one experience, then another. It stores traces as memory. It compares current perception with remembered perception. It anticipates possible outcomes. It uses sequence to survive.

This is necessary for consciousness.

But the universe may not be conscious in this narrative way.

The universe may not say, "That was my past, this is my present, that is my future." It may not hold memory like a human mind. It may not plan like a human being. It may not travel forward like a person walking along a road. It may simply operate through conditions and consequences.

Energy changes state.

Fields interact.

Disturbances seek resolution.

Structures form, sustain, transform, or dissolve.

Signals arrive after delay.

Observers arrange those consequences as time.

This is where Universal Energy Dynamics gives a different foundation.

Under UED, time is not denied, but it is repositioned. Time is not the first container of reality. It is the arrangement that appears when energy states resolve through causal sequence and alignment delay. The mind then narrates that arrangement as past, present, and future.

The mind needs time because it needs continuity.

But reality may need only consequence.

This difference is crucial.

For the mind, time is the stage. For the field, condition is the stage. For the mind, events happen in time. For UED, time appears because events take delay to become observable consequences. The mind experiences the delay and converts it into sequence.

This is why time feels so real. It is not because time is necessarily an independent cosmic road. It is because every conscious experience is filtered through change and memory. The mind cannot experience reality without organizing it. So the organizing tool becomes mistaken for the ultimate structure of reality.

A person wearing colored glasses for a lifetime may think the world itself has that color.

Similarly, the mind wearing time may think the universe itself is made of time.

This does not mean the mind is foolish. It means the mind is adaptive. Time is the mind's survival interface. It helps us distinguish cause from consequence, danger from safety, effort from result, memory from imagination, and action from outcome. Without time, practical life would fail.

But cosmology must not blindly project the mind's interface onto the universe.

This projection is exactly what creates the cosmic movie reel. Because the mind needs sequence, it imagines the universe as a sequence. Because the mind has memory, it imagines the universe has a stored past. Because the mind anticipates future, it imagines the universe moving toward a future. Because the mind understands growth through age, it expects galaxies to behave like living beings moving from childhood to maturity.

Then JWST sees mature-looking structures at cosmic depth, and the mind becomes surprised.

But perhaps the surprise is psychological before it is cosmological.

Perhaps the universe did not violate time.

Perhaps the mind overextended time.

The mind needs time to arrange experience. The universe may operate through field conditions resolving into consequence. Between those two, interpretation is born. If we confuse the mind's arrangement with the universe's foundation, we turn time into a cosmic road. Once that happens, every delayed signal becomes an old frame, every redshifted galaxy becomes early, every structure at depth becomes young, and every maturity becomes anomaly.

This is why we must understand why the mind needs time.

Only then can we stop mistaking that need for cosmic truth.

Time is the mind's way of holding change.

But the universe may be change resolving through consequence before the mind ever names it time.

10.2 — The Difference Between Duration and Cosmic Road

Duration is real.

A seed does not become a tree instantly. Water does not boil the moment it is placed on fire. A sound does not reach the ear before it travels through air. Light from the Sun does not arrive on Earth without delay. A child does not become an adult in one breath. A star does not form, shine, and collapse in the same instant.

Processes take intervals.

This cannot be denied.

But the existence of duration does not automatically prove that time is an independent cosmic road.

This distinction is very important.

In ordinary thinking, whenever something takes duration, we imagine that it is travelling through time. The seed travels through time to become a tree. The body travels through time to age. The universe travels through time to evolve. The galaxy travels through time to mature. This language feels natural because human experience is built on before and after.

But perhaps duration does not mean travel through a road.

Perhaps duration means resolution of a process.

Water takes time to boil because energy must enter the system, molecules must respond, internal motion must increase, and a threshold must be reached. The water is not travelling on a separate road called time. It is undergoing a change of state. The duration is the measurable gap between one condition and another.

A seed takes time to become a plant because conditions must interact: moisture, soil, temperature, nutrients, internal biological structure, sunlight, and growth response. The seed is not being carried by time like a passenger on a train. It is resolving through a chain of conditions. The duration is real, but it belongs to the process.

A sound takes time to reach the ear because air carries pressure disturbance at a finite speed. The delay is real, but it is not proof that sound is moving through an independent substance called time. It is proof that propagation requires medium, interaction, and sequence.

Similarly, light reaching a detector after delay does not automatically prove that the universe stores the past as a physical road behind us. It proves that signals require transmission. It proves that observation is not instantaneous. But the meaning of that delay must be interpreted carefully.

Duration is measurable.

The cosmic road is an assumption.

This is where modern cosmology often moves too fast. It sees signal delay and immediately builds cosmic history. It sees redshift and builds lookback time. It sees lookback time and builds early universe. It sees early universe and builds developmental expectation. Then, when observed structures do not obey that expectation, it calls them anomalous.

But the first fact was only duration.

The larger story was interpretation.

Universal Energy Dynamics does not deny duration. It gives duration a deeper foundation. Under UED, duration appears because energy states do not always resolve instantly. A field condition must interact, resist, align, saturate, or stabilize before a consequence becomes observable. The interval between condition and consequence is what we measure as time.

This is Alignment Threshold Time.

Time is not a road outside the process.

Time is the delay inside the process.

This one correction changes the entire cosmic imagination. If time is a road, then the universe is travelling from past to future. If time is delay, then the universe is resolving from condition to consequence. If time is a road, then distant light becomes an old frame of a cosmic film. If time is delay, then distant light becomes a signal whose arrival reflects propagation, resistance, alignment, and field condition.

The difference is subtle but revolutionary.

The cosmic road image makes time primary. Events happen inside time. Time exists first, and the universe moves within it. UED reverses this. Conditions exist. Interactions happen. Consequences resolve. Time appears as the measurable sequencing of this resolution.

In this view, the universe does not need time as a separate container.

It needs conditions.

It needs energy states.

It needs field interaction.

It needs resistance and alignment.

It needs consequence.

Time is the shadow of these processes, not their master.

This also explains why time feels different in different contexts. A minute of pain feels longer than a minute of joy. A year of stagnation feels different from a year of intense transformation. A process in one material occurs quickly, while the same input in another material produces a slow response. The clock may measure equal intervals, but the experience and resolution differ because the systems involved differ.

The clock gives comparison.

The process gives meaning.

Cosmology must learn this distinction. A cosmic clock may organize measurements, but it should not automatically dominate interpretation. If cosmic regions differ in field condition, then their resolution delays may differ. If signals carry resistance, then their observed timing may carry more than simple travel duration. If redshift includes field strain, then lookback time may not be the full meaning of distance.

This is why the idea of time as a cosmic road becomes dangerous.

It makes us overconfident. It allows us to place the universe on a single track. It allows us to say this object belongs to that age, this galaxy belongs to that childhood, this structure belongs to that stage. But if time is process-delay rather than a road, then such labels must become more careful.

A delayed signal is not automatically a full historical identity.

An old light reading is not automatically a primitive cosmic object.

A deep field observation is not automatically an early developmental stage.

Duration tells us that something took interval.

It does not alone tell us the final meaning of what is observed.

This distinction is also necessary for understanding Hubble tension. If time is a cosmic road, then expansion rate should fit neatly onto that road. But if time is alignment delay, then different field states may produce different apparent rates. What appears as disagreement in expansion may actually be disagreement in resolution delay. The problem may not be the universe's rate. The problem may be our assumption that all cosmic duration belongs to one uniform road.

Duration is real.

But duration belongs to processes.

The road is a mental image.

A useful image, yes. A powerful image, yes. But still an image.

The universe may not be moving along time.

The universe may be resolving through structured conditions, and the mind may be arranging those resolutions as time.

This is why we must separate duration from cosmic road. Without this separation, we will keep turning every delay into history, every history into a timeline, every timeline into expectation, and every violated expectation into anomaly.

The universe does not need to travel through time to change.

Change only needs conditions to resolve into consequence.

Time is what we call the delay between them.

10.3 — Causality Before Time

Before there is time, there is condition.

Before there is sequence, there is relation.

Before the mind says "this happened before that," reality has already produced a consequence from a cause. This is where our understanding of time must become deeper. We usually say that events happen in time, but perhaps events do not need time as an independent road. Perhaps events need conditions, interactions, resistance, and consequences. Time appears only when those consequences are arranged, measured, or remembered.

This is why causality may be more fundamental than time.

A seed does not sprout because time commands it to sprout. It sprouts because the right conditions act upon it: soil, moisture, temperature, internal structure, sunlight, nutrients, and biological readiness. If those conditions are absent, years may pass and nothing will happen. Time alone does not create the plant. Conditions do.

Water does not boil because time has passed. It boils because energy has entered the system, molecular activity has increased, pressure conditions allow transition, and the boiling threshold has been reached. If there is no heat, the water may sit for hours without boiling. Again, time alone does nothing. Causality acts. Time measures the duration of that action.

This is the key distinction.

Time does not cause events.

Conditions cause events.

Time records, measures, or narrates the interval between condition and consequence.

In Universal Energy Dynamics, this distinction becomes central. The universe is not moving through a pre-existing river of time. Energy states interact. Structured Vacuum Energy responds. Gradients arise. Resistance appears. Ripple behavior develops. Alignment occurs. Saturation may be reached. A consequence becomes observable. The mind or instrument then measures the delay and calls it time.

So time is not the first cause.

Causality is deeper.

Causality, in UED, is not a dry mechanical chain. It is the behavior of energy under conditions. It is the way one state produces another through interaction, imbalance, and resolution. Where energy is stable, little appears to happen. Where imbalance appears, action begins. Where action begins, consequence follows. Where consequence follows, sequence becomes meaningful. Where sequence becomes meaningful, time enters experience.

This is very different from saying that the universe exists inside time.

It suggests that time exists inside the interpretation of consequence.

When we say one event happened before another, we are describing an order of consequences. The order is real, but the order does not require time to be an independent cosmic substance. It requires causal relation. If one condition must exist before another can arise, the mind experiences this as before and after. But before and after are not necessarily cosmic places. They are relational positions in causal sequence.

Fire must touch dry grass before the grass burns.

A signal must be emitted before it is received.

An energy state must disturb the field before the field resolves.

A cause must act before a consequence appears.

This order is causality.

Time is how the observer arranges and measures that order.

The danger begins when we reverse the relationship. We start believing that time produces causality. We say, "As time passed, this happened." But time did not make it happen. Conditions made it happen. Time only measured the gap. If the conditions had not changed, the same time could have passed without the same consequence.

This is true in life, science, and cosmology.

A student does not become knowledgeable simply because two years pass. Learning happens only if attention, effort, guidance, reflection, and practice occur. A society does not progress simply because decades pass.

Progress happens only if institutions, ethics, policy, education, and collective intelligence align. A galaxy does not mature simply because cosmic time passes. Structure appears when field conditions allow condensation, coherence, regulation, and stability.

Time without causal change is empty duration.

Causality gives time meaning.

This is why modern cosmology may have overestimated the explanatory power of time. When deep galaxies appear mature, the standard question becomes, "How did this happen so early?" But this question assumes that chronological age is the main measure of possible structure. UED asks a deeper question: what conditions produced this structure?

If the conditions were present, structure could appear.

If the conditions were absent, even long time would not help.

This is why the idea of "too early" may be misleading. Too early according to what? According to a timeline. But if causality depends on field condition, then the real issue is not age alone. The real issue is whether the field conditions were sufficient for manifestation.

A mature-looking galaxy at cosmic depth may not be a violation of time.

It may be a consequence of field conditions that reached coherence.

A central regulator may not be a black hole that grew too fast.

It may be the saturation condition of a galactic field architecture.

A high redshift may not be only old light.

It may be the signature of field strain.

In each case, causality must be read before chronology.

This is where the cosmic movie reel begins to fail again. A movie reel depends on frame order. Frame one must look primitive, frame two slightly developed, frame three more complex, and so on. But the universe may not be a film advancing through fixed developmental stages. It may be a field in which conditions produce consequences wherever thresholds are reached.

The sequence exists.

But it is not necessarily the sequence imagined by the timeline.

The old model says: enough time must pass for complexity.

UED says: sufficient conditions must align for manifestation.

This does not remove time from science. It gives time its proper place. Time becomes the measurable delay between causal states. It becomes the interval through which alignment happens. It becomes the way observers compare processes. But time no longer becomes the ultimate creator of events.

The creator of events is condition acting through causality.

This also helps us understand why different systems show different rates. If time were a universal road, one might expect all processes to be placed equally on that road. But if time is the measured expression of causal resolution, then every system's timing depends on its condition. A dense field, a strained field, a saturated field, and a stable field will not resolve identically. Their causal delays will differ.

This is precisely why Alignment Threshold Time matters.

ATT is causality measured through field delay. It tells us that consequence does not appear merely because time passes. Consequence appears when energy states cross the threshold required for alignment. Until that threshold is reached, the process remains unresolved. Once it is reached, consequence becomes observable.

This means time is born from threshold behavior.

Not as a thing.

As a measure.

As an experience.

As narration.

The mind then takes these causal sequences and builds history. It remembers consequences and calls them past. It anticipates possible consequences and calls them future. It attends to current consequence and calls it present. This narration is useful and real to consciousness, but it should not be confused with the universe travelling on a road.

The universe may not be travelling.

It may be resolving.

This is the deeper meaning of causality before time. First there is field condition. Then there is interaction. Then there is resistance. Then there is alignment or saturation. Then there is consequence. Then there is measurement. Then there is narration. Time enters late in the chain, but the mind places it first because the mind experiences everything through sequence.

That reversal creates confusion.

It makes us think time causes change.

But change reveals time.

It makes us think the universe lives inside time.

But time may live inside the arrangement of universal consequences.

This is why cosmology must return to causality. Before asking how old an object is, we must ask what condition produced it. Before asking how early a galaxy formed, we must ask what field state allowed coherence. Before asking how fast the universe expands, we must ask what alignment delay is being measured. Before asking where the past is, we must ask how the mind arranges received consequences.

Causality is the real foundation.

Time is its measurable shadow.

10.4 — Alignment Threshold Time as Field Delay

Alignment Threshold Time is the delay between condition and consequence.

This is the simplest way to enter the idea.

A condition exists. Something changes within that condition. Energy enters, pressure builds, resistance appears, interaction begins, and the system does not immediately become a new state. It must pass through adjustment. It must reorganize. It must reach the point where the next state becomes possible.

That delay is not empty.

That delay is the process.

In Universal Energy Dynamics, this delay is called **Alignment Threshold Time**, or ATT. It is the measurable duration required for energy states to rearrange, align, and become observable as consequence. It is not time as a cosmic road. It is time as the field's delay in resolving change.

This idea is important because it allows us to understand time without turning it into a mysterious ruler standing outside reality.

Time is not sitting somewhere, waiting for events to move through it. Events do not travel along time like vehicles on a highway. Instead, energy states interact, resist, adjust, align, saturate, or dissolve. When this resolution takes duration, the observer measures that duration and calls it time.

Time is therefore the shadow of field delay.

Not the source of change.

Not the owner of causality.

Not the road beneath the universe.

A simple example can help. When water is heated, the water does not boil simply because minutes pass. It boils because energy has entered the system and the internal condition reaches the boiling threshold. If the heat is weak, the delay is longer. If the heat is strong, the delay is shorter. If pressure changes, the boiling point changes. The duration is real, but it depends on condition.

That is ATT in ordinary language.

The consequence appears only when the threshold is reached.

Before the threshold, the process is active but unresolved. After the threshold, the consequence becomes visible. Between the two lies delay.

Now take this principle from water to the field.

Structured Vacuum Energy is not passive emptiness. It carries conditions. It has resistance. It has tension. It has gradients. It allows ripple behavior, condensation, saturation, and resolution. When energy states interact within this field, consequences do not necessarily appear instantly. The field must resolve them according to its condition.

If the field is stable, the delay may be small.

If the field is strained, the delay may be different.

If the field is highly resistant, the delay may increase.

If the field is saturated, ordinary coherence may fail.

This is state-dependent time.

Not time changing because a clock is confused, but time appearing differently because the field condition requires different resolution delay.

This is why ATT becomes essential for cosmology.

Modern cosmology often treats time as a background dimension. The universe expands over time. Galaxies form over time. Light travels through time. Structures mature over time. The entire cosmic story depends on time as an independent ordering system.

But UED asks us to move deeper.

What if the "time" we assign to cosmic processes is actually the measurable delay of field resolution? What if cosmic rates are not pure movements along a universal road, but expressions of how different field conditions align into observable consequence? What if the universe does not age like a body, but resolves like a field?

This changes the interpretation of everything.

A galaxy appearing mature at cosmic depth is no longer automatically a violation of age. The deeper question becomes: what alignment threshold produced this structure? A redshifted signal is no longer only an old message. The deeper question becomes: what field delay and resistance shaped this signal? Hubble tension is no longer only a disagreement about expansion. The deeper question becomes: are different measurements reading different alignment delays?

ATT gives us the language to ask these questions.

It says that measurable time emerges from the resolution behavior of the field.

This also explains why different regions of the universe may produce different apparent rates. A locally stabilized field condition may resolve signals in one way. A deep, high-strain field condition may resolve differently. A compact region may carry strong resistance. A central saturation zone may prevent ordinary coherence from returning. If we measure all these through one uniform time model, disagreement becomes inevitable.

The disagreement is not necessarily in the universe.

It may be in the assumption that every field state must resolve through the same temporal reading.

This is precisely why Hubble tension becomes meaningful under UED. Local measurements may reflect one field-resolution condition. Deep-field or early-universe-based calculations may reflect another. If their alignment delays differ, their apparent expansion readings may differ. The standard model calls this tension. UED calls it state-dependent resolution delay.

In this view, time is not universal sameness.

Time is conditional delay.

This does not destroy measurement. It improves it. A clock still measures intervals. A telescope still detects signals. A model still estimates duration. But now the interpretation must include field condition. We cannot simply ask, "How much time passed?" We must also ask, "What condition was resolving during that interval?"

Because duration without condition is incomplete.

This is where ATT differs from ordinary clock time.

Clock time is comparison. It compares one regular process with another. A clock ticks, an atomic transition occurs, Earth rotates, a signal travels, and we compare intervals. This is useful. It gives order. It allows science, technology, communication, and prediction.

But ATT asks what lies beneath the interval.

It asks what process required the delay.

It asks what field resistance shaped the delay.

It asks what threshold had to be crossed before consequence became visible.

In this sense, clock time is surface measurement.

ATT is process meaning.

This does not make clocks wrong. It makes them partial. A clock tells us that duration occurred. ATT asks why that duration was necessary.

At cosmic scale, this distinction becomes decisive. If we only rely on clock-like interpretation, then the universe becomes a timeline. If we include ATT, the universe becomes a field of resolving conditions. The same observed delay may carry different meanings depending on resistance, tension, saturation, and alignment.

A delayed signal may not only be old.

It may be strained.

A slow process may not only be long.

It may be resistant.

A rate disagreement may not only be error.

It may be state-dependence.

This is why time as ATT is more powerful than time as road. A road is passive. It simply allows movement. But ATT is active because it belongs to the process itself. It tells us that change does not happen merely because time passes. Change becomes measurable when the condition resolves.

The universe, under UED, is not travelling along time.

It is resolving through thresholds.

Each consequence appears when the field condition permits it. Each signal arrives through delay shaped by resistance. Each structure becomes visible when energy aligns into form. Each saturation zone forms when coherence crosses its upper limit. Each observation is not merely placed in time; it is born through resolution.

This is why the cosmic movie reel fails.

A movie reel assumes a fixed sequence of frames. Frame one, frame two, frame three. Early, middle, late. Primitive, developing, mature. But a field does not need to obey such a simple sequence. A field resolves wherever conditions reach threshold. Structure may appear where alignment is ready. Saturation may appear where resistance is extreme. Signals may carry delay that is not reducible to chronological age.

ATT breaks the monopoly of the timeline.

It does not erase time.

It deepens it.

It moves time from the outside of reality to the inside of process. It removes time from the throne and places it where it belongs: between condition and consequence.

Time, then, is not a cosmic road.

It is the delay of the field becoming readable.

10.5 — Conscious Narration and the Birth of Past and Future

The past is not seen directly.

It is remembered.

The future is not seen directly.

It is imagined.

The present is not held permanently.

It is experienced as a moving interface between memory and expectation.

This is why time becomes so powerful inside consciousness. The mind does not merely receive events. It arranges them. It stores traces of what has happened, compares them with what is happening, and projects possible consequences ahead. From this arrangement, the mind creates three great zones of experience: past, present, and future.

But these zones may not exist in the universe the way they exist in the mind.

This is a difficult idea because human life depends on these zones. The past feels real because memory carries emotional weight. A childhood event can affect an adult decision. A failure can produce fear. A success can produce confidence. A loss can remain alive inside the mind for decades. The past may no longer be physically present, but it remains psychologically active.

The future also feels real because anticipation shapes action. A student studies for an exam that has not yet happened. A farmer prepares for rain that may come. A scientist designs an experiment for a result not yet known. A nation plans infrastructure for generations not yet born. The future is not physically present, but it influences present behavior.

The present is the only direct interface, yet even the present is unstable. The moment we name it, it begins to pass into memory. The mind catches reality in fragments, arranges those fragments into continuity, and says, "This is my life moving through time."

This is conscious narration.

The mind does not experience raw consequences as isolated flashes. It turns them into story. It says, "I was there. Then this happened. Because of that, I became this. Now I must do that." This narration gives identity, meaning, responsibility, fear, hope, guilt, and purpose.

Without narration, human life would lose coherence.

But cosmology must be careful not to project this narration onto the entire universe.

The universe may not possess a past in the way the mind possesses memory. It may not possess a future in the way the mind holds expectation. It may not move through a present in the way consciousness feels itself moving through attention. These may be observer-based arrangements of consequence, not independent cosmic regions.

Under Universal Energy Dynamics, the past and future are not denied as meaningful experiences. They are repositioned. The past is the mind's arrangement of resolved consequences. The future is the mind's projection of possible consequences. The present is the active interface where field consequence becomes observable to consciousness.

The field itself does not need narration.

It operates through condition, interaction, alignment, saturation, and consequence.

A signal arrives. A structure forms. A field state resolves. A body ages. A star changes. These are real processes. But when consciousness receives them, it arranges them into temporal identity. It says this came before, that came after, and something else may happen next.

This is why time feels like a road.

The road is built by narration.

Memory becomes the road behind us.

Expectation becomes the road ahead.

Attention becomes the place where we appear to stand.

But perhaps the universe is not walking on that road.

Perhaps only the conscious mind is drawing it.

This has deep implications for cosmology. When we say that JWST is looking into the past, we may be doing something similar. We receive delayed signals. We arrange them into history. We place objects into earlier chapters of the cosmic story. Then we judge them according to the developmental expectations of that story.

But what if the universe is not offering us a memory?

What if it is offering us a signal?

A delayed signal does not automatically mean the past exists as a physical archive "out there." It means a consequence has arrived through delay. The observer then arranges that arrival into a historical frame. This frame may be useful, but it may not be final.

The past, in cosmology, may be partly a narration built from received consequences.

The future, too, may be projection. We use current models to predict expansion, heat death, collapse, transformation, or cosmic fate. But these futures are not sitting ahead like destinations on a road. They are possible consequences inferred from present understanding.

The universe does not need to know its future.

The mind needs future to act.

This is why conscious narration is both powerful and dangerous. It gives meaning, but it can also create illusion. It helps us organize experience, but it may also force reality into a story it never agreed to follow. It allows us to build science, but it can also make us mistake model-history for cosmic truth.

Universal Energy Dynamics asks us to respect narration without worshipping it.

Memory is useful.

Prediction is useful.

History is useful.

But the universe may be deeper than memory, prediction, and history.

It may not be a film already recorded behind us and unwritten ahead of us. It may be a structured field where consequences resolve, and where conscious beings arrange those resolutions into the language of time.

This is why the past and future may be real to consciousness, but not necessarily independent regions of the universe.

The past lives as memory.

The future lives as possibility.

The present lives as observation.

Reality lives as consequence.

And time, as we experience it, may be the narration born when consciousness tries to hold that consequence in order.

10.6 — Why the Cosmic Movie Reel Fails

The cosmic movie reel fails because the universe may not be a film.

For generations, we have imagined the universe as if it were a grand sequence of frames. First there was the beginning. Then expansion. Then cooling. Then particles. Then atoms. Then stars. Then galaxies. Then planets. Then life. Then consciousness. Everything is arranged like a story moving from simplicity to complexity, from infancy to maturity, from past to present.

This arrangement is beautiful.

It is also deeply human.

The human mind understands reality through sequence. It cannot easily hold existence as a whole. It divides. It arranges. It says, "This came first, that came later." It turns change into chapters. It turns consequence into history. It turns delay into narrative. So when the universe was placed inside a similar structure, the mind found comfort.

The cosmic movie reel became the grand scientific story.

In that story, deep space became old light. Old light became past. Past became early universe. Early universe became cosmic childhood. Once this conversion happened, every distant signal was treated like a frame from an earlier part of the film. A galaxy at high redshift was no longer simply a structure with a shifted signal. It became an early galaxy. A compact red object became a primitive object. A central regulator became an early black hole. Everything was assigned a place in the cosmic film.

Then JWST disturbed the reel.

It looked into frames where the script expected infancy and found signs of maturity. It found brightness where faintness was expected. It found compactness where looseness was expected. It found structure where incompleteness was expected. It found possible central activity where slow growth was expected.

The movie reel began to stutter.

But perhaps the reel was always the wrong metaphor.

A delayed signal does not automatically mean a stored historical frame. A redshifted galaxy does not automatically mean a child of the universe. A distant observation does not automatically mean a primitive scene. These interpretations arise only when time is treated as a cosmic road and the universe is imagined as travelling through it frame by frame.

Universal Energy Dynamics offers a different foundation.

The universe is not primarily a film sequence. It is a structured field of consequences. Matter is ripple condensation within Structured Vacuum Energy. Gravity is energy-gradient behavior seeking balance. Black holes are Ripple Saturation Zones. Galactic centers are regulatory cores. Redshift carries field strain and ripple tension. Time is alignment delay and conscious narration.

Once all this is understood, the cosmic movie reel loses its authority.

The universe may not be playing old frames for us.

It may be resolving field states into observable consequences.

This changes the meaning of distance. A far signal is not only far in space; it may be field-deep. This changes the meaning of redshift. A red signal is not only old; it may be strained. This changes the meaning of maturity. A mature structure is not only old; it may be coherent. This changes the meaning of time. Time is not the road beneath the universe; it is the delay between condition and consequence.

The movie reel fails because it mistakes arrangement for reality.

It takes the mind's need for sequence and projects it onto the cosmos. It assumes that because consciousness organizes experience through past, present, and future, the universe itself must be arranged in the same way. But the universe may not need narration. It may not need chapters. It may not need a beginning-to-end film structure.

The field does not remember like a human mind.

It resolves.

The field does not plan like a human mind.

It responds.

The field does not travel through time like a character in a story.

It manifests consequences through condition.

This is why the cosmic movie reel produces anomalies. It forces the universe into developmental expectations. Early must be immature. Later must be mature. Distant must be young. Nearer must be older. Structure must grow slowly. Central regulators must grow after galaxies. Black holes must be endpoints. Redshift must be timestamp.

Then, whenever reality refuses these expectations, we call it strange.

But the strangeness belongs to the reel.

Not necessarily to the universe.

The movie reel also fails because it treats observation as passive recovery of history. It imagines telescopes as machines that collect old frames from the universe's archive. But observation may be more subtle. A signal becomes meaningful only when it resolves into measurement. It arrives through field conditions. It carries strain, resistance, delay, and transformation. The telescope does not simply open a drawer of cosmic memory. It participates in the present resolution of a field consequence.

This does not mean the past is useless.

It means the past must be understood properly.

The past is meaningful to consciousness. It is meaningful to science as reconstruction. It is meaningful as the arrangement of received consequences. But it may not exist as a physical region waiting behind us. It may not be a cosmic storage room. It may not be a film strip stretched across space.

When we say we are looking into the past, we must be careful.

We may be looking at delayed signals.

We may be looking through field depth.

We may be looking at strained consequences.

We may be looking at present resolution of distant interaction.

But we may not be looking at a dead frame in the simple way the movie-reel metaphor suggests.

This is why Chapter 10 is central to the entire book. Until time is reinterpreted, cosmic anomalies will continue to appear. If time remains a road, then the universe remains a film. If the universe remains a film, then every distant structure must obey its assigned frame. And if it refuses, the anomaly returns.

But once time becomes alignment delay and conscious narration, the universe opens differently.

The mature high-redshift galaxy is no longer necessarily in the wrong frame.

The Little Red Dot is no longer a premature object.

The central regulator is no longer too early.

The Hubble tension is no longer merely a disagreement in expansion.

They all become signs that field reality cannot be reduced to a simple temporal sequence.

The cosmic movie reel fails because reality is not obligated to be cinematic.

The universe may not be unfolding like a story written for observers. It may be a structured, equilibrium-seeking field where consequences become visible according to condition, not according to our preferred narration.

The mind creates the reel.

The universe creates consequences.

And when the reel cannot hold the consequences, we call them anomalies.

Perhaps it is time to stop repairing the reel.

Perhaps it is time to understand the field.

Closing Reflection — The Universe Does Not Travel Through Time

The universe may not be travelling through time.

This is a difficult sentence because the mind resists it immediately. Everything we know appears to happen in time. Bodies age. Stars change. Signals arrive after delay. Civilizations rise and fall. Memories collect behind us. Possibilities wait ahead of us. The entire human experience is arranged through the feeling that we are moving forward.

So when we say the universe does not travel through time, it sounds almost impossible.

But perhaps the impossibility belongs to habit.

The universe may not be moving along a road called time. It may be resolving conditions into consequences. Structured Vacuum Energy may be interacting, resisting, aligning, saturating, redistributing, and becoming measurable through events. The sequence we call time may arise because these consequences do not appear all at once. They emerge through delay, threshold, and resolution.

Time, then, is not the container of the universe.

Time is the measurable shadow of change.

This chapter has tried to separate what is useful from what is fundamental. Time is useful. It is necessary for memory, science, communication, planning, identity, and measurement. It helps consciousness arrange experience. It helps instruments compare processes. It helps language speak of before and after.

But usefulness does not mean independence.

A clock measures duration, but it does not prove that time is an invisible cosmic substance. A memory preserves the past in the mind, but it does not prove that the past exists as a physical region behind the universe. A prediction imagines the future, but it does not prove that the future is waiting ahead like a destination.

The mind needs time.

The universe may need only consequence.

This is the deepest shift.

In Universal Energy Dynamics, cause and consequence come before time. Conditions interact. Energy states disturb the field. Resistance appears. Alignment begins. Saturation may occur. A consequence becomes observable. Only after this does time become meaningful as sequence, delay, comparison, or narration.

This is why Alignment Threshold Time matters.

ATT tells us that measurable duration is not floating outside the process. It belongs to the process. It is the delay required for energy states to reach observable consequence. A stable field resolves one way. A strained field resolves another. A saturated field may fail to return ordinary coherence. Therefore, time cannot be separated from field condition.

This breaks the authority of the cosmic road.

If time were a universal road, then every cosmic event would simply occupy a place along that road. Galaxies would be young or old according to their frame. Redshift would become a timestamp. Distance would become history. JWST would become a machine looking backward through old scenes.

But if time is alignment delay and conscious narration, the universe becomes different.

A delayed signal is not automatically a dead historical frame. A redshifted galaxy is not automatically a cosmic child. A mature structure at high redshift is not automatically impossible. Hubble tension is not automatically a contradiction in expansion. These may all be signs that field reality has been compressed into a timeline too quickly.

The universe may not be a movie reel.

It may be a field of structured consequences.

The cosmic movie reel fails because it turns the mind's arrangement into cosmic law. It takes the human need for sequence and projects it onto existence. It assumes that because we experience life through past, present, and future, the universe itself must be arranged as past, present, and future in the same way.

But the field does not remember.

It resolves.

The field does not anticipate.

It responds.

The field does not move through time.

It expresses consequence.

This does not destroy history. It purifies it. History remains meaningful as reconstruction. Memory remains meaningful as experience. Scientific timelines remain useful as models. But they must no longer be mistaken for the ultimate structure of reality.

The past may be the mind's arrangement of resolved consequences.

The future may be the mind's projection of possible consequences.

The present may be the interface where consequence becomes observable.

Reality itself may be deeper than all three.

This is why cosmology must become careful with the word "early." Early according to what? Early according to signal delay? Redshift? Model-based expansion? Human narration? If time is not a road, then early cannot be treated as an unquestioned physical identity. It becomes an interpretive label.

And labels can mislead.

A galaxy is not young because we place it in an early frame. A central regulator is not premature because our timeline expected slower growth. A Little Red Dot is not impossible because it refuses the role of cosmic infancy. These objects and signals may be field consequences arriving through alignment delay and resistance.

The universe does not need to obey the ages we assign to it.

It only needs to operate through conditions.

This is the liberation offered by Chapter 10.

Once time is removed from the throne, the anomalies begin to lose their artificial drama. We no longer ask only, "How did this happen so early?" We ask, "What field condition produced this consequence?" We no longer ask only, "How far back are we looking?" We ask, "What delay, strain, and resistance shaped this signal?" We no longer ask only, "What was the universe like then?" We ask, "What is the field revealing now through this observation?"

The word now also changes here.

Now is not merely a clock moment. It is the point of observable resolution. It is where the signal meets the observer. It is where field consequence becomes measurement. It is where consciousness arranges the received event into meaning.

The universe may be still in its totality, yet active through consequences.

The mind experiences movement because it receives those consequences sequentially.

This prepares the next chapter.

If the universe does not travel through time, then perhaps motion itself must be re-examined. Perhaps the universe is not moving in the way the mind imagines. Perhaps what moves is not the totality, but the observer's arrangement of consequences. Perhaps the cosmos is fundamentally structured presence, while the mind experiences movement through change, contrast, and sequence.

That is where we must go next.

For now, this chapter leaves us with a quiet correction.

Time is real as experience.

Time is real as measurement.

Time is real as delay.

But time may not be the road beneath the universe.

It may be the shadow cast by consequences as they become observable.

And if that is true, then the universe is not travelling through time.

It is becoming readable through consequence.

Chapter 11 — The Still Universe and the Moving Mind

How the universe may be fundamentally still while the mind experiences motion, history, depth, and change through consequences.

When Stillness Looks Like Motion

The universe appears restless.

Everywhere we look, something seems to be moving. Planets orbit stars. Stars move within galaxies. Galaxies rotate. Light travels across vast cosmic depth. Particles interact. Fields fluctuate. Bodies age. Signals arrive. Structures form, transform, and dissolve. From the smallest interaction to the grandest cosmic structure, reality appears to be in motion.

So we naturally conclude that the universe itself is moving.

But perhaps this conclusion is too quick.

Motion is real within the universe, but that does not necessarily mean the universe as a whole is moving through something. A planet can move around a star because there is relation. A star can move within a galaxy because there is a larger field structure. A signal can travel because there is propagation across condition. But what does it mean to say that the universe itself is moving?

Moving where?

Through what?

Against which background?

A thing moves when its position changes relative to something else. But the universe, as totality, has no outside road. It has no external stage. It has no larger container through which it travels. If there is nothing outside the universe against which its movement can be measured, then perhaps the universe as a whole is not moving in the way our mind imagines.

Perhaps motion belongs to relations inside the universe.

Perhaps stillness belongs to the totality.

This is a difficult idea because the mind experiences everything through change. We do not encounter the universe as totality. We encounter fragments, signals, events, objects, delays, memories, and consequences. The mind receives one thing after another. It compares what was with what is. It imagines what may be. From this sequence, it creates the feeling of flow.

The mind moves through meaning.

Reality may be resolving through consequence.

This is where Universal Energy Dynamics gives us a deeper lens. Under UED, the universe is not primarily a machine travelling through time. It is a structured field of energy conditions. Matter is ripple condensation. Gravity is energy-gradient equilibrium. Black holes are Ripple Saturation Zones. Redshift is field strain. Time is alignment delay. What we call motion is often the visible result of imbalance seeking resolution within Structured Vacuum Energy.

In this view, a moving planet is not merely a body travelling through empty space. It is part of a field-gradient relationship. A rotating galaxy is not merely a collection of stars moving around a center. It is a field architecture maintaining coherence. A redshifted signal is not merely old light moving from the past. It is a field consequence resolving through resistance and alignment delay.

Motion remains real.

But its meaning changes.

It is not proof that the universe is travelling along a cosmic road. It is proof that field conditions are resolving into observable consequences.

The mind, however, cannot experience all consequences at once. It receives reality in sequence. It remembers previous consequences and calls them past. It anticipates possible consequences and calls them future. It attends to present consequence and calls it now. This creates the feeling that existence is flowing forward.

But perhaps the flow is not in the universe as totality.

Perhaps the flow is in consciousness.

A still lake can contain ripples. The lake as a whole may remain present, while ripples move across its surface. A screen can remain still while images move upon it. A field can remain continuous while local disturbances arise, resolve, and disappear. These analogies are limited, but they help us enter the idea: local motion does not automatically mean total movement.

The universe may be fundamentally still as totality, while active in its consequences.

This does not make the universe lifeless. On the contrary, it makes it more profound. It suggests that stillness is not absence of activity. Stillness is the wholeness within which activity becomes possible. The field does not need to travel anywhere to produce consequences. It only needs conditions, interactions, resistance, alignment, and resolution.

The mind sees change and says, "The universe is moving."

UED sees consequence and asks, "What field condition is resolving?"

This distinction matters because cosmology has often taken the mind's experience of flow and projected it onto the cosmos. It has imagined the universe as a story moving from beginning to future, as a film running through time, as a grand journey from origin to destiny. Then every distant signal becomes an earlier scene, every redshift becomes an old frame, every mature deep structure becomes a contradiction.

But if the universe is fundamentally still, and only consequences are resolving, then the movie reel loses its authority.

We may not be watching the universe move through time.

We may be watching field conditions become readable.

We may not be looking backward into a dead past.

We may be resolving deeper layers of a structured present.

We may not be seeing cosmic history as stored film.

We may be seeing consequences arranged by the moving mind.

This chapter begins from that quiet possibility.

The universe may not be rushing anywhere.

It may not be travelling through time.

It may not be unfolding like a movie made for observers.

It may be still in its totality, while the mind experiences motion through the sequence of consequences it receives.

Perhaps motion is not the final truth.

Perhaps motion is how stillness becomes readable.

11.1 — The Difference Between Local Motion and Total Stillness

A planet moves.

A star moves.

A galaxy rotates.

Light travels.

Signals arrive.

Nothing in this chapter denies motion at the local level. To deny local motion would be absurd. We observe motion everywhere. We measure it, calculate it, predict it, and depend on it. The entire structure of physics, astronomy, engineering, and daily life relies on the reality of movement.

But the deeper question is not whether things move within the universe.

The deeper question is whether the universe itself moves as a thing.

This is where the distinction becomes important.

A planet moves around a star because there is a relational structure. The planet has a changing position relative to the star. A star moves within a galaxy because there is a larger galactic field in which its motion can be described. A galaxy moves relative to other galaxies because observational relations allow comparison. Motion requires reference. It requires contrast. It requires relation.

Without relation, motion has no meaning.

If a single object existed in absolute emptiness, with no background, no observer, no other object, and no measurable reference, what would it mean to say that it was moving? Moving relative to what? Moving through what? Moving against which field of comparison?

This question becomes even deeper when we speak of the universe as totality.

If the universe includes all that exists physically, then it has no outside stage. It has no external road. It has no larger container through which it travels. It cannot move in the same way a planet moves, because a planet moves within a larger system. The universe has no larger physical system around it in which its total motion can be measured.

This does not mean nothing happens.

It means totality is not a travelling object.

Local structures move within the universe. But the universe itself, as totality, may be fundamentally still.

This is not stillness as deadness.

It is stillness as completeness.

A lake may remain present as a whole while ripples move across its surface. A screen may remain still while images move across it. A musical instrument may remain in place while vibrations arise through it. These examples are not perfect, but they reveal the difference between local activity and total movement.

The whole does not need to travel for activity to appear within it.

Universal Energy Dynamics pushes this idea further. Under UED, the universe is not primarily a collection of objects moving through empty space. It is a structured field of energy conditions. Motion appears when energy gradients, ripple disturbances, resistance, and alignment processes resolve into observable consequences. What moves is not the universe as totality, but relations within the field.

This changes the meaning of motion.

A falling stone is not merely an object travelling through space. It is a condensed field structure moving according to a local gradient condition. A planet's orbit is not merely a body circling through emptiness. It is a dynamic balance within an energy-gradient relationship. A rotating galaxy is not merely stars circling a center. It is a large-scale field architecture maintaining coherence around a regulatory core.

Motion is real.

But it is relational.

It belongs to conditions inside the field.

The universe as totality does not need to move through time or space for these relations to occur. The field only needs difference, imbalance, resistance, and resolution. Once a condition differs from another condition, relational behavior appears. Once relational behavior appears, motion becomes measurable. Once motion is measured, the mind arranges it into time, history, and change.

But the total field remains the total field.

This distinction helps us avoid a common mistake. Because we see motion everywhere inside the universe, we imagine the universe itself must be moving somewhere. Because galaxies appear to recede, we imagine the whole cosmos as travelling through an expanding road. Because light arrives after delay, we imagine the universe as moving from past to future like a film. But these are interpretations built from local relations.

The totality may not be travelling.

It may be expressing.

It may be resolving.

It may be becoming readable through local consequences.

This is why local motion and total stillness can coexist.

There is no contradiction.

A body can have internal motion while remaining whole. A field can contain disturbances while remaining continuous. A system can display dynamic relations without itself being a travelling object. Similarly, the universe can contain motion, formation, dissolution, radiation, rotation, and signal propagation while still being fundamentally still as totality.

The mind struggles with this because it lives inside sequence. It does not experience the whole universe at once. It experiences one consequence after another. It sees the Sun rise, the day pass, the night come, the body age, the memory grow, the signal arrive. From this, the mind concludes that everything is flowing forward.

But flow may be the mind's experience of sequential consequence.

It may not be the movement of the universe as a whole.

This becomes very important for cosmology. If we confuse local motion with total cosmic travel, then the universe becomes a story moving through time. We begin to speak as if the cosmos itself has a childhood, youth, maturity, and future destiny in the same way a living organism does. We begin to treat distant light as old frames from that story. We begin to judge deep galaxies by where they appear in the film.

But if the universe is still as totality, this whole framing becomes questionable.

The universe may not be moving from one cosmic chapter to another.

Instead, field conditions may be resolving into observable consequences. Signals may be arriving through alignment delay. Structures may be appearing through ripple condensation. Redshift may be showing field strain. Gravity may be showing energy gradients seeking balance. The mind then arranges all this into a moving story.

The motion is real inside the story.

But the story may not be the final reality.

This is the subtle power of total stillness.

It does not deny movement. It places movement correctly. Motion belongs to relation. Relation belongs to field condition. Field condition belongs to totality. Totality itself does not need an outside path.

The universe may not be a traveller.

It may be the field within which all travel becomes meaningful.

11.2 — Motion as Consequence, Not Cosmic Journey

Motion is usually imagined as travel.

Something is here. Then it is there. A body changes position. A planet moves around a star. A galaxy rotates. A signal crosses space. The mind sees this change of position and calls it motion.

This is correct at the surface level.

But Universal Energy Dynamics asks us to look deeper.

Motion may not be merely the movement of objects through empty space. Motion may be the visible consequence of field imbalance seeking resolution. What appears as travel may actually be energy states adjusting through Structured Vacuum Energy. What appears as movement may be the field resolving difference, tension, resistance, and alignment.

This changes the entire meaning of motion.

In ordinary thinking, an object moves as if it is separate from the space around it. A stone falls through air. A planet travels through space. A photon travels across the universe. A galaxy moves away from another galaxy. The object is treated as the actor, and space is treated as the passive stage.

But under UED, the stage is not passive.

The vacuum is not empty.

The field participates.

If matter is ripple condensation within Structured Vacuum Energy, then every object is already a field consequence. It is not a separate thing moving inside a dead container. It is a local coherence within a continuous field. When that coherence shifts, responds, aligns, or follows a gradient, we observe motion.

So motion is not object-first.

Motion is field-first.

A falling stone does not fall simply because Earth pulls it like a hand pulling a rope. It moves because Earth and stone exist within a field-gradient relationship. Earth represents a massive condensation of energy within Structured Vacuum Energy. The stone is another condensed field structure. The field between and around them carries gradient difference. The motion we call falling is the consequence of that gradient resolving.

In the old language, gravity pulls.

In the UED language, imbalance guides.

The same applies to orbit. A planet orbiting a star is not simply a small body trapped by a larger body. It is a sustained field-gradient relationship. Its motion is neither pure escape nor pure fall. It is dynamic balance. The planet continues its path because motion and gradient remain in relation. The orbit is not merely travel around a center. It is a visible expression of equilibrium under tension.

A galaxy's rotation can also be read this way.

In the ordinary view, stars move around a galactic center, held by gravity and supported by unseen matter when visible matter seems insufficient. In UED, the galaxy is not a pile of stars moving independently. It is a field architecture. Its rotation is the consequence of large-scale gradient behavior, ripple condensation, central regulation, and coherence within Structured Vacuum Energy.

The stars move, yes.

But their motion belongs to a larger field pattern.

This is why motion should not be interpreted as cosmic journey too quickly.

A local object may move. A signal may propagate. A structure may rotate. But these motions do not prove that the universe as totality is travelling somewhere. They only prove that field relations are active. They prove that energy differences are resolving. They prove that consequences are becoming observable.

Motion is consequence.

Not necessarily journey.

This distinction becomes especially important when we speak about cosmic expansion. Galaxies appear to recede. Redshift is measured. The standard interpretation says space itself is expanding, and the universe is moving through its expansion history. The mind then imagines a grand cosmic journey: from beginning to present, from compact past to expanded now, from early simplicity to mature structure.

But UED asks whether this may be a misreading of field behavior.

What if apparent expansion is not the universe travelling outward through a cosmic road, but the large-scale observational expression of energy redistribution within Structured Vacuum Energy? What if redshift is not merely the stretching of space, but the resistance signature of ripple behavior across field depth? What if the apparent motion of galaxies is partly the way field gradients become readable at cosmic scale?

Then motion is no longer a simple journey through time.

It becomes the field expressing resolution.

This also helps us understand why motion and stillness are not opposites in the deepest sense. A field can remain whole while local conditions change within it. A still ocean can contain waves. A silent atmosphere can carry sound. A stable body can contain heartbeat, circulation, electrical activity, and cellular motion. The whole can remain coherent while parts move within it.

Similarly, the universe may be still as totality while consequences move within it.

This does not make motion unreal. It makes motion relational. It says that motion is not an independent proof of cosmic travel. It is the measurable expression of changing relationships inside the field.

In this sense, every motion is a message.

A falling object tells us about local gradient.

An orbit tells us about dynamic balance.

A rotating galaxy tells us about field architecture.

A redshifted signal tells us about resistance and alignment delay.

A jet near a central regulator tells us about energy redistribution around saturation.

Motion reveals condition.

It does not automatically reveal a cosmic road.

The mind, however, is tempted to turn motion into story. It sees change and says, "The universe is going somewhere." It sees sequence and says, "Reality is moving forward." It sees delayed light and says, "We are looking backward." It sees redshift and says, "The universe is expanding through time."

But perhaps the mind is narrating what the field is resolving.

The universe may not be on a journey.

The field may be in consequence.

This is why the idea of motion must be refined before we complete the end of the cosmic movie reel. If motion is understood only as travel through space and time, the movie reel survives. Every object is a character. Every signal is an old scene. Every redshift is a timestamp. Every galaxy is a frame in the cosmic film.

But if motion is understood as field consequence, the reel weakens.

The universe is no longer a story moving from one scene to another. It is a structured totality whose local conditions resolve into observable behavior. The observer then arranges that behavior as movement, history, distance, and change.

This also brings us closer to the role of consciousness.

The mind experiences motion because consequences arrive in sequence. It cannot absorb all conditions simultaneously. It receives signals, stores memory, compares states, and narrates flow. Motion becomes not only a physical relation but also an experiential arrangement. The field resolves. The mind experiences movement.

This is the bridge between cosmology and consciousness.

The universe may be still as totality.

The field may be active in consequence.

The mind may experience that activity as motion.

Therefore, motion is not false.

It is simply not final.

It is the readable surface of deeper field resolution.

A universe understood this way is not dead. It is not frozen. It is not static in the ordinary sense. It is dynamic through consequence while still as totality. It does not need to travel anywhere to express motion. It does not need an external road to generate change. It does not need time as a highway to reveal process.

It only needs field condition.

It only needs imbalance.

It only needs resistance.

It only needs alignment.

And from these, motion appears.

Motion is the universe becoming readable through consequence, not the universe travelling through a road outside itself.

11.3 — Why the Mind Experiences a Flow

The mind experiences reality as flow because the mind cannot receive everything at once.

It receives in sequence.

A sound comes, then a thought follows. A sight appears, then a memory awakens. A feeling rises, then a response forms. The mind does not experience all consequences simultaneously as one completed whole. It receives fragments, arranges them, connects them, and creates continuity.

This continuity feels like flow.

We say time is passing. We say life is moving. We say the world is changing. We say we are going forward. But beneath these expressions, something more subtle is happening: the mind is organizing received consequences into a moving experience.

The flow may not be in the universe as totality.

The flow may be in consciousness.

This does not mean the flow is fake. It is real as experience. A flowing river is real to the one standing beside it. Aging is real to the body. Memory is real to the mind. Change is real to life. But the felt movement of time may be the way consciousness processes consequence, not proof that the universe itself is travelling along a cosmic road.

The mind is built to detect change.

It notices difference. It remembers one state and compares it with another. It sees the Sun higher than before and says morning has moved toward afternoon. It sees a child taller than before and says years have passed. It sees an old photograph and says the past is gone. It sees a plan and says the future is coming.

The mind does not merely observe change.

It narrates change.

This narration creates the feeling of a flowing reality. The past appears behind, the future appears ahead, and the present feels like a moving point between them. But this arrangement belongs to the mind's interface. It is the way consciousness stabilizes experience.

Without such flow, life would become impossible.

If the mind could not arrange events, learning would fail. If it could not remember consequences, responsibility would weaken. If it could not anticipate outcomes, survival would become difficult. The mind needs flow because action requires direction. A person must know that fire burned yesterday, food is needed today, and tomorrow may bring hunger again. Sequence protects life.

So the mind creates a living timeline.

This timeline is practical, but it may not be cosmic.

Universal Energy Dynamics asks us to distinguish between the mind's flow and the field's resolution. The field does not need to narrate. It does not need to say, "I was this, now I am that, later I will become something else." It operates through condition and consequence. Energy states interact. Resistance appears. Alignment begins. Saturation may occur. Consequence becomes observable.

The mind receives these consequences one by one and experiences movement.

This is why the same duration can feel different in different states of mind. One hour of joy passes quickly. One hour of pain feels long. A waiting mind stretches time. An absorbed mind forgets time. A fearful mind accelerates thought. A peaceful mind slows experience. The clock may show equal intervals, but the flow inside consciousness changes.

This tells us something important.

The flow of time is not only physical measurement.

It is also mental processing.

The mind does not merely read time; it manufactures the felt texture of time through attention, memory, anticipation, and emotional state. The same external interval can become heavy or light depending on the inner condition of the observer.

This does not make external processes meaningless. It simply shows that experienced time is not identical with mechanical duration. Consciousness adds narration to sequence. It gives emotional weight to delay. It converts neutral change into personal history.

Cosmology has made a similar move at a larger scale.

It receives delayed signals and arranges them into cosmic history. It receives redshift and arranges it into lookback time. It receives structure at depth and places it into the early universe. It receives brightness or compactness and compares it against the expected frame. Then, if the object refuses the assigned stage, it becomes anomalous.

But perhaps this is the cosmic version of the mind's narration.

The universe offers consequences.

The mind builds the reel.

This is why the flow of cosmic history must be questioned. The flow may not belong to the universe in the same way it belongs to human consciousness. It may belong to the interpretive arrangement through which observers convert signals into sequence.

The field may be resolving.

The mind may be narrating.

The difference is enormous.

In UED, the universe is a structured field of energy conditions. Motion, redshift, gravity, matter, and signal delay are all expressions of field behavior. The mind experiences these expressions in sequence and creates a sense of flow. It does not see the total field. It receives partial resolutions. It does not hold all consequences at once. It arranges them as before and after.

This is why we feel that reality is moving.

The mind is always comparing.

A comparison requires at least two states. The moment the mind compares, sequence is born. The moment sequence is born, time appears. The moment time appears, flow begins. The mind then mistakes this flow for the movement of the universe itself.

But the universe may not be flowing.

It may be present as totality, while local consequences become readable through relation.

This is not easy to grasp because human consciousness is inside the process. We cannot step outside experience and observe the universe from a timeless external place. We can only reason from within sequence. That is why time feels unavoidable. The observer is part of the narration.

Still, intelligence can separate experience from ultimate interpretation.

We can accept that the mind experiences flow while also asking whether the universe itself flows. We can accept that motion is real locally while questioning whether totality moves. We can accept that history is meaningful while questioning whether the past exists as a cosmic archive.

This is the maturity required now.

The mind experiences flow because it is not the whole field. It is a receiver of consequences within the field. It is not total reality; it is an interface. It receives, stores, compares, predicts, and narrates. Through this activity, it converts consequence into experience.

And experience becomes flow.

This also explains why stillness is so difficult for the mind. Perfect stillness would remove contrast. Without contrast, there is no sequence. Without sequence, there is no felt movement. Without movement, the mind loses its usual way of knowing. The mind depends on difference to remain active. It needs change to maintain identity.

If nothing changed, the mind would have nothing to compare.

If nothing compared, time would lose meaning.

This is why the mind turns even stillness into thought. It imagines, remembers, anticipates, questions, and narrates. Even when the external world becomes quiet, inner movement continues. The mind keeps producing flow because its own functioning depends on sequential processing.

The universe may not need such processing.

The universe does not need to remember itself to exist.

It does not need to anticipate itself to become.

It does not need to narrate itself to resolve.

It simply operates through field conditions and consequences.

The mind, however, needs narration to understand what has resolved.

This is why cosmic flow may be partly an observer effect—not in the shallow sense that the observer invents the universe, but in the deeper sense that the observer arranges received consequences into temporal experience. The field provides the consequence. The mind provides the flow.

This distinction helps dissolve the cosmic movie reel.

The movie reel exists because the mind needs frames. It needs beginning, middle, and end. It needs old and new. It needs early and late. It needs before and after. But field reality may not be a reel. It may be a continuous structured presence in which local consequences become observable according to condition.

The mind experiences those consequences as movement.

The field remains what it is.

Therefore, the flow of time may be the mind's experience of sequential consequence.

Not the universe rushing somewhere.

Not reality travelling along an invisible road.

Not the cosmos playing a film for observers.

The field resolves.

The mind flows.

And between them, time is born as experience.

11.4 — History as Arranged Consequence

History is not the past itself.

History is the arrangement of consequences after they have resolved.

This distinction may look small, but it is enormous. The past does not walk into the present as a complete object. It does not sit before us like a physical structure waiting to be touched. What we call the past arrives through traces: memory, records, fossils, light, ruins, documents, effects, scars, patterns, and consequences. The mind then arranges these traces into a story and calls that story history.

History is meaningful.

But history is arranged.

It is not raw reality exactly as it was. It is reality reconstructed through consequences available now. A historian studies documents, events, causes, records, and outcomes. A scientist studies signals, fossils, radiation, and traces. A person studies memory, emotion, and experience. In every case, the past is known through what has remained, what has arrived, what has been interpreted, and what has been arranged.

This is why history is powerful but incomplete.

It gives meaning to what happened. It connects events. It explains how one condition led to another. It helps societies learn. It helps individuals understand themselves. It helps science reconstruct processes that are no longer directly present.

But history should not be mistaken for the past as a living region.

The past is not physically standing behind us like a place. It is not a warehouse where events are stored in their original form. It is not a film archive waiting to be replayed. What exists now are consequences, traces, signals, memories, and records. The mind arranges them into order.

This is true in human life.

A person remembers childhood, but childhood is not physically present. It exists through memory, photographs, habits, wounds, strengths, stories, and consequences carried into the present. The child no longer exists as the original event. But the consequences of childhood remain active in the adult's body, mind, choices, fears, confidence, and relationships.

So when the adult says, "This happened in my past," the statement is meaningful. But what is actually present now is the consequence of what happened.

The past is reconstructed through present traces.

The same logic applies to civilization. A nation remembers war, reform, migration, suffering, achievement, revolution, and decline. But those events are not physically sitting somewhere behind the present. They exist through documents, monuments, laws, wounds, institutions, inherited attitudes, and social consequences. History is the arrangement of these consequences into a meaningful sequence.

Now apply this to cosmology.

When we say that telescopes look into the past, we must be careful. A telescope does not enter a physical room called the past. It receives a signal. That signal arrives now. The detector records it now. The data is processed now. The model interprets it now. The mind arranges it into cosmic history now.

The signal may be delayed.

But the history is arranged.

This does not make the signal false. It does not make cosmology meaningless. It simply means that cosmic history, like all history, is built from present consequences. The light arriving from a distant structure is a consequence. Redshift is a consequence. Brightness is a consequence. Spectrum is a consequence. The model then arranges these consequences into a timeline.

The danger begins when arrangement is treated as direct reality.

A high-redshift signal is received. It is placed into the early universe. The structure is assigned an age. That age becomes developmental identity. Then the object is judged according to the expected stage of that history. If it appears mature, it becomes anomalous.

But perhaps the anomaly is born inside the arrangement.

The universe gave us a signal.

We gave it a biography.

Universal Energy Dynamics asks us to separate these layers. The signal is a field consequence. It may carry delay, strain, resistance, ripple tension, alignment history, and observational transformation. Before we turn it into a fixed historical frame, we must ask what field condition produced it and what field condition shaped its arrival.

History is useful after consequence is understood.

But when history is imposed too early, it becomes a trap.

This is what happened with the cosmic movie reel. The universe was arranged into a grand developmental sequence. Then every distant observation was forced into that sequence. High redshift became early history. Early history became cosmic childhood. Cosmic childhood became expected immaturity. Then JWST revealed structures that did not behave like children, and the model became uncomfortable.

But perhaps the discomfort is not proof that the universe has a strange past.

Perhaps it is proof that our arrangement of cosmic history is too rigid.

History, in UED, is not denied. It is repositioned. It becomes the mind's ordered reading of resolved consequences. It becomes a useful model, not a final container. It helps us understand sequence, but it must not replace mechanism.

The mechanism is field condition.

Energy states interact. Structured Vacuum Energy responds. Ripple behavior appears. Matter condenses. Gravity emerges as gradient behavior. Saturation zones regulate. Redshift carries strain. Signals arrive through alignment delay. Then the observer arranges these consequences into history.

This means history comes late.

Reality first produces consequence.

History arranges consequence afterward.

This is why cosmic history must remain humble. It is not the universe speaking in final chapters. It is our best reconstruction from available signals. As signals improve, the reconstruction must change. JWST has done exactly that. It has not broken the universe. It has challenged the way cosmic consequences were arranged into history.

A mature high-redshift galaxy does not necessarily mean the early universe was impossible. It may mean the historical placement of that signal was incomplete. A Little Red Dot does not necessarily mean structure formed too quickly in the past. It may mean compact field condensation was misread as primitive chronology. A central regulator at cosmic depth does not necessarily mean a black hole grew impossibly fast. It may mean the galaxy and regulator arose as one field architecture, later arranged into a growth story.

History is not wrong.

But history is not enough.

It must be guided by mechanism.

Without mechanism, history becomes narration. With incomplete mechanism, history becomes misleading narration. With field-based mechanism, history becomes a more disciplined reconstruction of consequence.

This is why the still universe and moving mind matter. The field resolves conditions. The mind receives consequences in sequence. Then the mind arranges them into history. The movement of history belongs partly to the mind's need for order. The universe itself may not be storing past frames. It may be presenting consequences that become historical only through interpretation.

This does not reduce wonder.

It increases responsibility.

Every time we call something ancient, early, late, mature, premature, or impossible, we must ask: are we describing the universe, or are we describing our arrangement of its consequences?

History is necessary.

But it must remain transparent.

It must admit that it is built from traces. It must admit that signals arrive through field conditions. It must admit that the past is not held directly but reconstructed through present evidence. It must admit that when new evidence arrives, the arrangement may need to change.

This is the maturity required in cosmology.

We must stop treating cosmic history as a rigid film reel. We must treat it as a reconstruction built from delayed, strained, field-marked consequences. Only then can we avoid turning every mismatch into an anomaly.

History is the mind's ordering of consequence.

The universe gives consequence.

The mind creates history.

And when the mind forgets this, history becomes a cage.

11.5 — Depth Without Pastness

Depth does not always mean past.

This is one of the hardest habits to break in cosmology because the habit has become almost automatic. The deeper we look into space, the farther we believe we are looking. The farther we look, the older the signal must be. The older the signal, the earlier the universe must be. The earlier the universe, the more primitive the structure should appear.

This chain feels logical.

But it may be incomplete.

Depth can mean distance, yes. It can involve delay, yes. It can carry lookback relation, yes. But depth may also mean something more fundamental: layered field condition. A signal may not be deep only because it has travelled across a large spatial separation. It may be deep because it has emerged through a field state very different from our local observational condition.

This is the UED shift.

Under Universal Energy Dynamics, the vacuum is not empty space. It is Structured Vacuum Energy. It is the field condition within which matter condenses, gravity appears, redshift emerges, signals propagate, and observations become measurable. Therefore, looking deeper into the universe may not mean only looking farther back into chronological history. It may also mean resolving deeper layers of field resistance, ripple tension, energy density, and alignment delay.

This changes the meaning of deep observation.

A telescope may not be merely an instrument of pastness.

It may be an instrument of field penetration.

When JWST observes high-redshift galaxies, compact red objects, and possible central regulators, the standard interpretation quickly places them into the early universe. Once placed there, they are judged by the expectations of earlyness. They should be young, faint, loose, immature, and still assembling.

But what if JWST is not only looking backward?

What if it is looking deeper?

What if the high redshift does not only say, "This object belongs to the ancient past," but also says, "This signal has emerged through a field condition of high resistance and strain"?

Then the observation changes meaning.

A mature structure at cosmic depth may not be a contradiction. It may be a deeper field architecture. A Little Red Dot may not be a premature object in cosmic infancy. It may be a compact ripple-condensation zone. A central black-hole-like region may not be a monster that grew too fast. It may be a Central Ripple Saturation Zone arising with the galactic field itself.

The anomaly appears only when depth is forced into pastness.

This is not a denial of signal delay. A distant signal does not arrive instantly. Delay is real. Transmission is real. The question is whether delay alone should control the entire meaning of the observation. A delayed signal tells us that the signal has taken duration to reach us. It does not automatically tell us that the object must be interpreted as an immature historical stage.

That second step is interpretation.

And interpretation must remain open.

In human life, depth also does not always mean past. When we say a person has deep understanding, we do not mean the person is old. We mean the person has entered a deeper layer of meaning. When we say a problem has deep roots, we do not only mean it began long ago. We mean it has layers beneath the surface. When we say a relationship is deep, we do not mean it is ancient only. We mean it carries intensity, connection, and complexity.

Depth means layered reality.

Cosmic depth may need the same maturity of meaning.

It may include distance, but not only distance.

It may include delay, but not only delay.

It may include history, but not only history.

It may include field state.

This is where the cosmic movie reel becomes weak. A movie reel converts depth into frame order. The farther frame belongs to earlier time. The earlier frame must show earlier development. If something in that frame looks advanced, the story becomes disturbed. But a field is not a movie reel. A field has layers. A deeper layer may carry more tension, more resistance, more density, and more compact structure. It need not be primitive merely because it is deep.

The ocean analogy helps again. When we look deeper into the ocean, we do not say we are looking into the ocean's past. We are looking into a different layer of pressure, darkness, temperature, density, and life. The deep ocean is not the young ocean. It is the deeper condition of the same ocean.

Similarly, cosmic depth may not always mean cosmic youth.

It may mean deeper field condition within the same universal reality.

This does not mean all cosmological distances are illusions. It means distance is not the only meaning of depth. Spatial separation, signal delay, redshift, field resistance, and alignment threshold may all participate in what we call deep observation. If we reduce all of them to pastness, we flatten the universe into a timeline and lose the field.

Depth without pastness allows the universe to become layered again.

It allows high-redshift structures to be studied without immediately forcing them into childhood. It allows redshift to carry field strain. It allows brightness and compactness to belong to energetic coherence. It allows galactic centers to be understood as regulators instead of late-forming monsters. It allows JWST to become not only a time machine, but a field-depth instrument.

This is why the phrase "looking back" must be replaced with caution.

Perhaps sometimes we are looking back.

But perhaps we are also looking through.

Looking through field resistance.

Looking through ripple tension.

Looking through alignment delay.

Looking through the layered structure of the present universe.

The mind wants to turn depth into past because past is easy to arrange. It gives comfort. It gives sequence. It gives a beginning and a developmental order. But the universe may be more subtle than that. It may not be arranged mainly for the convenience of our narration. It may reveal itself through conditions that do not fit neatly into early and late.

This is why mature-looking deep galaxies matter so much.

They may not be telling us that the early universe was impossible.

They may be telling us that depth was misunderstood.

They may be telling us that cosmic observation is not merely a backward journey through time, but a deeper reading of field reality. Their maturity may not be oldness. It may be coherence. Their compactness may not be premature growth. It may be condensation. Their redshift may not be pure distance. It may be field strain.

Once this is understood, the universe no longer appears as a broken historical sequence.

It appears as layered presence.

The distant does not automatically become primitive.

The deep does not automatically become old.

The red does not automatically become ancient.

The visible does not automatically become final.

A signal from cosmic depth may be a present consequence of a deeper field condition arriving through delay. It may become historical only when the mind arranges it into a timeline. That arrangement may be useful, but it should not be treated as final reality.

Depth is real.

Pastness is interpretation.

And when we confuse the two, the universe begins to look anomalous.

But perhaps the anomaly disappears when we allow depth to be depth before forcing it to become history.

11.6 — Change Without Cosmic Travel

Change is real.

This must be said clearly before any deeper interpretation begins. The universe is not frozen in the ordinary sense. Stars ignite and die. Galaxies rotate and merge. Planets form, cool, fracture, and transform. Bodies age. Minds learn. Signals arrive. Structures appear and disappear. Nothing in this discussion denies change.

But change does not automatically mean the universe is travelling through time.

This is the difference we must hold carefully.

In the ordinary imagination, change means movement along a road. Something was one way before, becomes another way now, and may become something else later. From this, the mind concludes that reality is travelling through time. The universe becomes a traveller, and time becomes the road beneath it.

But Universal Energy Dynamics asks us to look deeper.

Change may not require a road.

Change may require only condition, interaction, resistance, alignment, and consequence.

Water changes into steam when energy conditions reach a threshold. A seed changes into a plant when biological and environmental conditions align. A star changes through stages because energy, pressure, density, and field conditions evolve through consequence. A galaxy changes because field gradients, ripple condensation, central regulation, and energy redistribution continue to interact.

In each case, change is not caused by time itself.

Change is caused by conditions resolving.

Time measures the delay. The mind narrates the sequence. But the real driver is consequence.

This is why UED does not deny change. It removes the false throne given to time. The universe does not need an external dimension called time to transform. It needs structured energy states interacting within Structured Vacuum Energy. Once imbalance appears, the field responds. Once resistance appears, alignment begins. Once threshold is crossed, consequence becomes observable.

That observable shift is called change.

The mind then says time has passed.

But perhaps what really passed was not time.

Perhaps a condition resolved.

This distinction becomes extremely important in cosmology. If change is understood as travel through time, then the universe becomes a story moving from beginning to future. Galaxies must be young before mature.

Structures must be primitive before complex. Distant light must belong to earlier frames. High-redshift objects must belong to cosmic childhood.

But if change is consequence without cosmic travel, then the interpretation opens.

A galaxy appearing mature at cosmic depth may not be a structure that travelled too quickly through time. It may be a field architecture whose conditions reached coherence. A Little Red Dot may not be a premature object. It may be compact field condensation. A central regulator may not be a black hole that grew too fast. It may be a saturation boundary arising with the galactic field itself.

The question changes from "how did this happen so early?" to "what field condition produced this change?"

That is a better question.

Because time alone never explains transformation.

If a field condition is not ready, duration will not produce structure. If alignment threshold is not crossed, consequence will not appear. If resistance remains unresolved, stability will not emerge. If energy does not condense coherently, matter will not manifest as form. Time may pass in measurement, but the required change will not happen.

This is visible even in ordinary life. A person does not become wise merely because years pass. A student does not understand a subject merely because lectures continue. A society does not become ethical merely because decades move forward. Change requires causal condition. Without the right condition, time becomes empty waiting.

The same principle must be applied to the cosmos.

The universe does not become structured merely because cosmic time passes.

Structure appears when field conditions allow consequence.

This is why the idea of cosmic travel becomes misleading. It makes us think the universe moves from one stage to another because time carries it forward. But perhaps the universe is not moving from stage to stage. Perhaps field conditions are resolving into different observable states. The stages are our arrangement. The consequences are real.

A star's life is not denied by this. It has sequence. It has process. It has measurable duration. But that sequence is not proof that time is an independent cosmic road. It is proof that stellar conditions transform through causal resolution.

A galaxy's evolution is not denied. Galaxies may interact, merge, redistribute energy, form stars, and reorganize. But these changes belong to field processes, not to an external journey of the universe through a time-road.

The universe changes through consequences.

Not by travelling through something outside itself.

This also helps explain the stillness discussed in this chapter. Total stillness does not mean absence of local change. It means the universe as totality does not need to move as a thing. Local conditions can transform within totality. Local structures can move, interact, and resolve. Signals can propagate. Minds can experience sequence. But the total field is not travelling through an external container.

The whole remains.

The relations change.

The field resolves.

The mind experiences motion.

This is change without cosmic travel.

It is not a denial of dynamic reality. It is a correction of where dynamism belongs. Dynamism belongs to relations within the field. It belongs to energy gradients, ripple interactions, alignment thresholds, and consequence. It does not require the universe as totality to be a moving object.

This is why the cosmic movie reel fails again. A movie reel requires the universe to be a sequence of frames moving through time. Change is interpreted as frame progression. But a field reality does not need frames. It needs conditions. When conditions resolve, change becomes observable. The mind later arranges those changes into frames.

The reel is built afterward.

The field acts first.

This correction also changes the meaning of cosmic history. If the universe is not travelling through time, then history is not a physical film stored behind us. History is the mind's arrangement of resolved consequences. It may be meaningful and useful, but it is not the final structure of reality. The universe does not need to preserve its old frames. It only needs to carry the consequences of its field conditions.

This is why JWST's observations are so important.

They may be showing us change without the old timeline. They may be revealing field structures that our cosmic movie reel placed too early. They may be exposing the difference between causal consequence and chronological expectation. They may be forcing us to admit that maturity is not always age, depth is not always pastness, and change is not always travel through time.

The universe may be transforming locally while remaining still in totality.

That is not contradiction.

It is field logic.

A wave can move across water while the body of water remains the medium. A thought can move through the mind while awareness remains the field of experience. A pattern can change on a screen while the screen itself remains present. Again, these analogies are limited, but they show the same principle: local change does not require total travel.

Under UED, change is the field becoming readable through difference.

Difference creates contrast. Contrast allows observation. Observation allows sequence. Sequence allows narration. Narration becomes time. But the original reality is field condition resolving into consequence.

This means change is not an illusion.

Change is the language of consequence.

What is illusory is the belief that change proves an independent cosmic road.

The universe does not need to go somewhere to change. It does not need to move into the future. It does not need to carry its past behind it. It does not need to travel through a timeline. It only needs to express consequences from its conditions.

The mind experiences those consequences as motion, history, aging, development, and flow.

That experience is real for the mind.

But the deeper universe may be still as totality, active as field, and readable through change.

This is the balance we need.

We must not deny motion.

We must not deny change.

We must not deny history.

But we must stop treating them as proof that the universe is travelling through time like a character in a story.

Change does not prove cosmic travel.

Change proves consequence.

And once this is understood, the still universe and moving mind are no longer opposites. They are two levels of one reality: the field resolving, and consciousness experiencing resolution as movement.

Closing Reflection — The Mind Moves Through What the Field Resolves

Perhaps the universe is not moving as much as we think.

Perhaps it is resolving.

This is not an easy distinction, because the mind lives inside movement. We wake, work, remember, worry, hope, age, learn, and wait. Everything feels like motion. Life feels like a river. Thought feels like flow. History feels like a road. The universe itself appears to be expanding, evolving, and travelling from origin toward destiny.

So naturally, we imagine the cosmos as a moving story.

But perhaps that story belongs more to the moving mind than to the universe as totality.

The universe may be fundamentally still, not in the sense of being inactive, frozen, or dead, but in the sense of being complete as totality. It has no outside road to travel. It has no external stage against which its total movement can be measured. It has no larger container through which it moves. Motion exists within it, but the whole itself may not move like an object.

This distinction changes everything.

Planets orbit. Stars burn. Galaxies rotate. Signals travel. Bodies age. Structures form and dissolve. All this is real. But these are local consequences within the field. They do not prove that the universe itself is travelling through time like a character in a story.

Under Universal Energy Dynamics, the field resolves conditions. Energy gradients seek balance. Ripple disturbances align or saturate. Matter condenses into form. Gravity appears as field-gradient behavior. Redshift carries strain and resistance. Signals arrive through alignment delay. Structures become visible when field conditions reach coherence.

The field resolves.

The mind receives.

The mind arranges.

And from this arrangement, movement becomes experience.

This is why the moving mind is so important. The mind does not receive the whole universe at once. It receives fragments. It receives signals, sensations, memories, comparisons, and consequences. It cannot hold totality directly. It must arrange what arrives. It must create continuity. It must say, "This happened, then that happened, and now this is happening."

That arrangement becomes flow.

The mind calls it time.

The mind calls it history.

The mind calls it change.

But perhaps the deeper universe is not flowing in the same way. Perhaps what flows is the conscious narration of resolved consequences. The field does not need narration. It does not need memory. It does not need anticipation. It does not need to call one state past and another future. It simply operates through condition and consequence.

This does not make human experience false.

It makes human experience local.

The mind's flow is real to the mind. Aging is real. Memory is real. Delay is real. History is meaningful. Motion is measurable. But their reality does not automatically prove that the universe as totality is moving through an independent road called time.

The universe may not be travelling.

It may be becoming readable.

This is a profound correction.

When a signal reaches a telescope, we call it old light. But the observation happens now. The detection happens now. The interpretation happens now. The signal becomes meaningful now, at the point where field consequence meets instrument and consciousness. We may arrange it into history, and that arrangement may be useful, but the signal itself is a resolved consequence arriving through delay, strain, and field condition.

When JWST observes deep galaxies, perhaps it is not simply opening old frames from the universe's childhood. Perhaps it is resolving deeper layers of the present field. Perhaps the structures that appear mature are not breaking the timeline. Perhaps the timeline itself is too narrow to hold field-depth reality.

This is why the cosmic movie reel fails.

A movie reel belongs to the mind. It requires frames. It requires before and after. It requires a beginning, middle, and end. It requires objects to appear in assigned stages. It requires distant to mean early, early to mean immature, and mature to mean late.

But the universe may not obey cinema.

It may obey consequence.

A galaxy is not forced to be young because its signal is delayed. A central regulator is not forced to be premature because our model assigned it to an early frame. A Little Red Dot is not forced to be impossible because its compactness appears in a field layer we misread as cosmic infancy.

These may be field consequences, not narrative mistakes by the universe.

The mind moves through what the field resolves. That is the key line.

The field produces consequence. The mind experiences sequence. The field offers signals. The mind builds history. The field carries depth. The mind converts depth into pastness. The field displays change. The mind turns change into journey.

This is not weakness. This is how consciousness survives. The mind must arrange reality to act within it. It needs time, history, identity, expectation, and direction. Without them, practical life would collapse.

But cosmology must not confuse the survival interface of the mind with the deepest structure of the universe.

The still universe and moving mind are not enemies. They are two levels of understanding. At one level, motion is real, measurable, and necessary. At another level, totality does not travel. It contains relations through which motion becomes meaningful.

This is why stillness should not be misunderstood as absence.

Stillness is not emptiness.

Stillness is not inactivity.

Stillness is not death.

Stillness is wholeness.

The field can be whole and still produce ripples. The universe can be total and still express local change. Reality can be complete and still become readable through consequences. The mind sees those consequences one after another and experiences movement.

Perhaps this is the deepest reconciliation.

The universe is still as totality.

The field is active through conditions.

The mind moves through consequence.

This also changes the meaning of cosmic depth. We may not be looking backward in the simple way we imagined. We may be looking through field layers. We may be resolving strained signals. We may be encountering consequences from different field conditions. The sense of pastness may arise when the mind arranges delayed signals into history.

But depth does not always mean past.

Change does not always mean cosmic travel.

Motion does not always mean the universe is going somewhere.

The universe may be present beyond the mind's timeline.

And perhaps the mind, moving through its own narration, mistakes its movement for the movement of reality itself.

This chapter leaves us with that quiet possibility.

The cosmos may not be rushing forward.

It may not be carrying its past behind it.

It may not be walking toward a future destination.

It may be a structured field of consequences, still in its totality and active in its expressions.

The mind receives these expressions and moves through their meaning.

That movement is life.

That movement is history.

That movement is experience.

But the deeper field remains.

And if we understand this, then many cosmic anomalies begin to lose their drama. They were not necessarily failures of the universe. They were failures of the moving mind to understand the still field.

Perhaps the universe is not moving away from us.

Perhaps we are moving through the meanings of what the universe resolves.

Perhaps we are not looking backward into a dead archive.

Perhaps we are learning to read the depth of a present field.

And perhaps the end of cosmic anomalies begins when the moving mind finally learns humility before the still universe.

Chapter 12 — The End of Cosmic Anomalies

Why galaxies, black holes, redshift, cosmic maturity, and Hubble tension cease to be anomalies when seen through Universal Energy Dynamics.

The Universe Was Never Broken

The universe was never broken.

It only appeared broken because we were reading it through a frame that had become too narrow for what observation was beginning to reveal.

This is the quiet conclusion toward which the entire book has been moving. At first, the cosmos seemed to be filling with anomalies. Mature galaxies appeared too early. Little Red Dots appeared too compact and intense. Black holes appeared as impossible monsters of infinite density. Redshift seemed to stretch the timeline into discomfort. Hubble tension refused to settle into one clean expansion rate. JWST looked deeper and the old story became restless.

Naturally, the language of crisis began.

Too massive.

Too bright.

Too early.

Too mature.

Too fast.

Too strange.

Too impossible.

But the universe did not say these words.

We did.

A galaxy does not know that we have called it early. A redshifted signal does not know that we have converted it into ancient history. A dark galactic center does not know that we have named it a black hole and imagined a singularity inside it. A disagreement between measurements does not know that we have called it crisis.

Reality simply presents consequence.

The model decides whether the consequence is comfortable.

This is the heart of the matter.

An anomaly is not merely something unusual. It is something that refuses the expectation of a framework. The object itself may be perfectly natural. The signal may be meaningful. The structure may be coherent. The darkness may be functional. The tension may be informative. But if the framework is too rigid, the natural begins to look impossible.

This book has not tried to deny observation.

It has tried to rescue observation from inherited interpretation.

The galaxies are there. The redshift is measured. The dark centers influence motion. The Hubble tension persists. JWST has revealed structures that demand explanation. None of this should be dismissed. A serious framework must not escape data by calling it illusion.

But a serious framework must also not worship old interpretation when data begins to resist it.

Universal Energy Dynamics enters at this point.

It does not say the universe is wrong. It says the reading may be incomplete. It does not say matter does not exist. It says matter is not the beginning. It does not say gravity is unreal. It says gravity may be energy-gradient equilibrium. It does not say black-hole-like phenomena are imaginary. It says the monster of singularity may be interpretive. It does not say redshift is false. It says redshift may carry field strain and resistance, not only distance and pastness. It does not say time is meaningless. It says time may be alignment delay and conscious narration, not an independent cosmic road.

Once this lens is applied, the anomalies begin to change shape.

The mature galaxy is no longer necessarily an object that grew too fast. It may be a field architecture whose coherence was misread as chronological maturity.

The Little Red Dot is no longer necessarily a tiny impossibility in the early universe. It may be a compact ripple-condensation zone expressing high field tension.

The black hole is no longer necessarily a cosmic monster with infinite density. It may be a Ripple Saturation Zone where ordinary coherence fails to return.

The event horizon is no longer necessarily a magical wall. It may be the boundary of recoverable signal coherence.

The galactic center is no longer merely a destructive endpoint. It may be a central regulator holding the field pattern.

Redshift is no longer merely the universe's timestamp. It may be the color of resistance across Structured Vacuum Energy.

Hubble tension is no longer necessarily a collapse of cosmology. It may be evidence that different field states resolve through different alignment delays.

Time is no longer the road beneath the universe. It may be the shadow of consequence as it becomes observable.

And once all these are seen together, the cosmos begins to look whole again.

Not simple.

Not fully solved.

Not emptied of mystery.

But whole.

This is important. The end of cosmic anomalies does not mean the end of scientific inquiry. It does not mean there is nothing more to discover. It does not mean every number is settled, every mechanism is complete, and every observation has been explained forever. That would be another form of arrogance.

The end of anomalies means something subtler.

It means the contradiction weakens when the frame changes.

The universe no longer looks like a failed timeline. It begins to look like a layered field of structured consequence. What appeared as impossible under a matter-first, time-first, movie-reel model begins to appear meaningful under a field-first, consequence-first, UED model.

This is the final integration of the book.

The cosmos may not be a slow construction project of dead matter moving through empty space. It may not be a film strip of old frames stored across distance. It may not be governed by a single cosmic clock. It may not be filled with monsters, patches, exceptions, and impossible objects.

It may be a structured field where energy condenses, gradients seek balance, saturation regulates coherence, signals carry strain, and observers arrange consequences into time and history.

If we read it wrongly, it becomes strange.

If we read it deeply, it becomes coherent.

This is why the universe did not become strange overnight. It was our instruments that became stronger. It was our data that became sharper. It was our assumptions that became exposed. JWST did not break the cosmos. It broke the comfort of old interpretation.

That is not a disaster.

That is progress.

Every great shift in knowledge begins like this. First, reality refuses expectation. Then the defenders of the old frame call the refusal an exception. Then exceptions multiply. Then patches become heavier. Then someone asks whether the problem is not in reality, but in the frame.

This book stands at that question.

What if galaxies are not early in the way we think?

What if black holes are not holes?

What if redshift is not only distance?

What if time is not a road?

What if Hubble tension is not crisis?

What if the universe is not broken?

The answer offered here is not a final closure. It is a new opening.

The cosmos was never begging us to repair it.

It was inviting us to reinterpret it.

And when interpretation changes, the word anomaly begins to lose its power.

12.1 — What Makes Something an Anomaly?

An anomaly is not born in reality first.

It is born at the point where reality meets expectation.

This is the simplest way to understand the word. Something appears. A signal is measured. A structure is observed. A behavior is detected. Reality presents itself. Then the model looks at what has appeared and asks: does this fit what I expected?

If it fits, we call it confirmation.

If it does not fit, we call it anomaly.

But the object itself has not changed.

The galaxy has not become strange because our timeline is uncomfortable. The redshift has not become rebellious because our interpretation is incomplete. The dark center of a galaxy has not become a monster because our mathematics reaches infinity. The Hubble tension has not become cosmic failure because our expansion readings do not agree.

Reality simply appears as consequence.

The anomaly appears inside interpretation.

This is not to say anomalies are meaningless. They are extremely important. In fact, anomalies are among the most valuable gifts in science. They show where a model is under pressure. They reveal where expectation has become too narrow. They expose the difference between what reality is doing and what theory allows reality to do.

But we must be very clear: an anomaly is not automatically a defect in the universe.

It is a mismatch between observation and frame.

This distinction is the foundation of the final chapter.

A mature-looking galaxy at high redshift is not anomalous by itself. It is only anomalous if the model has already decided that such a galaxy must belong to the early universe and must therefore be immature. Without that expectation, the galaxy is simply a structure revealing brightness, compactness, mass, coherence, and field condition.

A Little Red Dot is not anomalous by itself. It is only anomalous if we expect compact red structures at cosmic depth to behave like weak primitive objects. Without that expectation, it may be read as a compact ripple-condensation zone, a high-tension field structure, a concentrated signature of energy alignment.

A black-hole-like region is not anomalous by itself. It is only anomalous if we force it into the image of infinite density, zero volume, singularity, and cosmic law-breakdown. Without that interpretation, it may be understood as a Ripple Saturation Zone, a field condition where coherence becomes unrecoverable to the external observer.

Hubble tension is not anomalous by itself. It is only anomalous if the universe is expected to produce one clean universal expansion rate through one smooth cosmic timeline. Without that assumption, the disagreement may reveal state-dependent resolution delay.

An anomaly, then, is not a thing.

It is a relationship.

It is the relationship between observation and expectation.

This is why anomalies can disappear without observations disappearing. When the framework changes, the same observation may no longer look strange. It may become meaningful. The data remains, but the discomfort changes.

This has happened many times in the history of knowledge. An observed fact that once looked impossible under one framework becomes natural under another. The fact did not wait for permission. It existed. The model had to mature.

The same may be happening now in cosmology.

The current crisis of anomalies may not mean the cosmos is breaking. It may mean the older interpretive frame is reaching its limit. That frame begins with matter as independent substance, space as empty background, gravity as attraction or curved fabric, redshift as distance-clock, time as cosmic road, and the universe as a movie reel moving from origin to maturity.

Inside that frame, many things become difficult.

Mature galaxies at cosmic depth become too early.

Compact red structures become too intense.

Central regulators become too massive too soon.

Black holes become impossible singularities.

Redshift becomes overloaded.

Hubble tension becomes crisis.

But perhaps the difficulty is not scattered.

Perhaps all these anomalies are symptoms of one deeper mismatch.

The universe may be field-first, while the model is matter-first.

The universe may be consequence-first, while the model is timeline-first.

The universe may be structured through field conditions, while the model reads it as historical frames.

This is why Universal Energy Dynamics changes the meaning of anomaly.

UED does not deny the observations. It does not remove the galaxy, the redshift, the dark center, the tension, or the signal. It changes the frame through which these are understood. Matter becomes ripple condensation. Gravity becomes energy-gradient equilibrium. Black holes become Ripple Saturation Zones. Galactic centers become regulatory cores. Redshift becomes field strain and resistance. Time becomes alignment delay and conscious narration.

Once this happens, the anomaly categories begin to dissolve.

Not because the universe becomes simple.

But because the interpretation becomes deeper.

This is a crucial difference. A shallow framework tries to remove anomaly by forcing data back into old expectations. It adds patches. It adjusts parameters. It invents exceptions. It protects the original story as long as possible. Sometimes this is useful, but after a point, too many patches become a sign that the frame itself needs revision.

A deeper framework does not fear anomaly.

It listens to anomaly.

It asks: what assumption is being exposed?

That is the correct scientific attitude.

When JWST reveals mature-looking structures at cosmic depth, the question should not only be, "How did this fit into the old timeline?" The question should also be, "Is the timeline the right lens?"

When Hubble tension persists, the question should not only be, "Which number is wrong?" The question should also be, "Are both numbers reading different field states?"

When black holes lead to singularities, the question should not only be, "What happens at infinite density?" The question should also be, "Does physical infinity indicate a real object or a failure of model coherence?"

When redshift produces discomfort, the question should not only be, "How far back are we looking?" The question should also be, "What field resistance has shaped this signal?"

This is how anomalies become gateways.

They are not enemies of science.

They are invitations to think beyond the frame.

The danger lies in treating the frame as sacred. Once a model becomes emotionally protected, every anomaly becomes a threat. But if truth is more important than comfort, every anomaly becomes a teacher. It tells us where reality is asking for better language.

The universe is not responsible for fitting our model.

The model is responsible for understanding the universe.

This is why the end of cosmic anomalies does not mean the end of mystery. Mystery remains. Measurement remains. Testing remains. Refinement remains. But the artificial contradiction weakens when the lens changes. The universe stops looking like a broken machine and begins to look like a layered field of consequences.

An anomaly is often the universe saying:

Your observation is real.

Your frame is incomplete.

Look again.

Read deeper.

Change the lens.

That is the spirit of this final chapter. The end of cosmic anomalies begins when we stop accusing reality and start examining interpretation. The galaxy was not too early. The black hole was not necessarily a hole. The redshift was not merely a timestamp. The Hubble tension was not necessarily cosmic confusion.

They became anomalies because the old frame made them so.

And once the frame changes, the cosmos begins to speak in a different language.

12.2 — Galaxies Without the Slow Assembly Trap

The slow assembly trap begins with a simple belief.

A galaxy must be built slowly.

Matter must gather. Gas must cool. Stars must form. Smaller structures must merge. Gravity must organize. Dark matter must provide scaffolding. Time must do its long work. Only after enough duration can a galaxy become bright, massive, compact, organized, and mature.

This belief feels reasonable because human experience supports it.

A house is built slowly. A city grows slowly. A forest expands slowly. A civilization matures slowly. Complexity, in ordinary life, appears to require time. So when we look at the cosmos, we naturally carry the same expectation. A young universe should produce young structures. A mature galaxy should require an old cosmic stage.

But this is precisely where the trap begins.

The slow assembly model turns time into the main builder of galaxies. Once time becomes the builder, every mature-looking structure at cosmic depth becomes suspicious. If JWST observes a bright compact galaxy at high redshift, the model asks: how did it assemble so quickly? If a Little Red Dot appears intense and dense, the model asks: how did such compact energy become organized so early? If central activity appears in deep-field systems, the model asks: how did the central black hole grow fast enough?

The same question keeps returning in different forms.

How did this happen so early?

But perhaps the question is already trapped.

It assumes that the galaxy must be a matter-collection requiring long chronological growth. It assumes that maturity means age. It assumes that compactness requires slow accumulation. It assumes that structure must first be loose, then gradually become organized. It assumes that the universe builds galaxies like humans build cities.

Universal Energy Dynamics breaks this trap by changing the foundation.

A galaxy is not merely a pile of collected matter.

A galaxy is a field architecture.

Matter itself is ripple condensation within Structured Vacuum Energy. Mass is a field-density signature. Gravity is energy-gradient equilibrium. The galactic center is a regulatory saturation zone. Rotation, luminosity, compactness, and structure are not separate puzzles added later; they are connected expressions of the same field condition.

Once this lens is applied, maturity changes meaning.

Cosmic maturity does not have to mean chronological oldness alone. It can mean field coherence. A galaxy may appear mature because its field architecture is coherent, not because it has travelled through a long mechanical childhood. Its compactness may indicate concentrated ripple condensation. Its brightness may indicate energetic expression from field alignment. Its central regulator may indicate inward saturation necessary for structural balance.

The galaxy is no longer judged only by age.

It is read through condition.

This is why high-redshift mature-looking galaxies cease to be impossible under UED. They may not be structures that somehow cheated the timeline. They may be regions where Structured Vacuum Energy reached the threshold required for coherent manifestation. Once field conditions align, structure can appear with surprising authority. The process is not magical. It is threshold-based.

A cloud forms when conditions allow condensation.

A whirlpool appears when flow conditions create rotation.

A crystal forms when the medium reaches the right state for order.

A galaxy, under UED, may appear when the field reaches the right state for large-scale ripple condensation and gradient coherence.

This does not mean galaxies do not evolve. They do. Stars may form, die, redistribute energy, and enrich local structures. Galaxies may interact, merge, stretch, distort, stabilize, and transform. Later evolution remains real. But foundational coherence may not require the same slow mechanical construction imagined by matter-first cosmology.

A galaxy may be born as architecture before every internal detail is finished.

This distinction is important.

A house under construction needs bricks placed one by one. But a field pattern can appear as a whole relational structure when conditions cross threshold. Its later details may continue developing, but the basic pattern is already present. Under UED, a galaxy is closer to a field pattern than a pile of bricks.

This is why Little Red Dots become so significant.

In the old model, they look troubling because they are compact, red, bright, and intense in regions assigned to the early universe. They appear too concentrated for the expected slow assembly story. They force the model to search for accelerated star formation, unusual stellar populations, dense gas, active nuclei, dust effects, or hidden black hole activity.

Some of these factors may exist in individual cases.

But UED asks a deeper question.

What if Little Red Dots are compact high-tension field condensation zones?

Their small appearance may not mean weakness. Their redness may not only mean distance. Their brightness may not only mean accelerated stellar history. Their compactness may not be a problem of impossible accumulation. Together, these features may indicate field strain, ripple tension, resistance, and localized saturation within Structured Vacuum Energy.

In that reading, the Little Red Dot is not a premature galaxy.

It is a concentrated field consequence.

The old framework separates its features and turns each into a problem. Redshift becomes age. Compactness becomes excessive density. Brightness becomes impossible luminosity. Maturity becomes chronological contradiction. UED reunites the features as one field-state expression.

The object is not rebelling against the universe.

It is revealing the field.

This is the larger correction for galaxies. The slow assembly trap exists because the old model begins with material parts and asks how long those parts needed to gather. UED begins with field condition and asks what threshold allowed structure to become visible. The first question produces the crisis of early maturity. The second question allows cosmic maturity to be understood as coherence.

A mature galaxy is not necessarily an old galaxy.

It may be a coherent galaxy.

A compact galaxy is not necessarily a galaxy that accumulated too much matter too quickly.

It may be a field-condensed structure.

A bright deep-field galaxy is not necessarily an impossible early object.

It may be an energetic expression of intense alignment.

This does not remove the need for observation, measurement, and caution. Every galaxy must still be studied. Mass estimates must be refined. Spectra must be analyzed. Distances must be tested. Luminous sources must be understood. UED does not give permission for careless certainty.

But it changes the meaning of surprise.

Instead of treating mature deep galaxies as violations of the cosmic timeline, it treats them as clues that the timeline may not be the deepest framework. Instead of asking only how galaxies assembled so fast, it asks whether assembly is the correct primary language. Instead of using age as the master measure, it uses field coherence, condensation, gradient balance, and regulatory structure.

This is why galaxies cease to be anomalies under UED.

They are no longer forced to behave like children in a human biography of the universe. They are no longer judged by a slow mechanical ladder from primitive to mature. They are understood as structured field consequences appearing when conditions reach threshold.

The universe did not create galaxies too early.

The old model called them early too quickly.

The universe did not produce impossible maturity.

The old frame confused coherence with age.

The galaxy was never trapped.

Only our interpretation was.

12.3 — Black Holes Without the Monster Myth

The black hole became an anomaly because it became a monster.

Not only in public imagination, but also in interpretation.

A dark region was inferred. Extreme gravitational behavior was observed. Stars moved around invisible compact centers. Energetic activity appeared near galactic cores. Light failed to return in ordinary ways. Mathematics moved toward collapse. The event horizon became a boundary. The singularity became the imagined center.

Then the monster was born.

A place where nothing escapes.

A point of infinite density.

A hole in the universe.

A region where physics breaks.

A cosmic mouth swallowing matter, light, and time.

But perhaps the monster was never in the universe.

Perhaps the monster was in the interpretation.

Universal Energy Dynamics does not deny the observed phenomenon. It does not deny dark compact regions. It does not deny extreme gravitational effects. It does not deny event-horizon-like behavior. It does not deny that galactic centers contain powerful invisible regulatory conditions. Something real is there. Something powerful is happening.

But the question is: what is it?

The traditional answer says it is a black hole, a collapsed object so dense that nothing can escape, with a singularity at its center. But UED asks whether this answer carries too many inherited assumptions. If matter is not an independent substance, but ripple condensation within Structured Vacuum Energy, then matter cannot simply be imagined as falling endlessly into an impossible point. If mass is field-density signature, then extreme mass behavior may indicate extreme field condition, not infinite material possession. If gravity is energy-gradient equilibrium, then extreme gravitational behavior may indicate extreme gradient saturation, not a monster-force crushing reality into absurdity.

Once the foundations change, the black hole changes.

It becomes a **Ripple Saturation Zone**.

This is the central correction.

A Ripple Saturation Zone is a region where Structured Vacuum Energy reaches such intense field condition that ordinary ripple coherence cannot return to the external observer in readable form. The region appears dark not because reality has ended, but because signal recovery has failed. It appears extreme not because an infinite-density object sits at the center, but because energy-gradient intensity has crossed the threshold of ordinary coherence.

This removes the need for singularity.

It does not weaken the phenomenon. It makes the phenomenon more physically meaningful.

Singularity is often treated as if it were a final scientific truth, but it may be closer to a warning sign. When a theory produces infinite density and law-breakdown, perhaps nature is not telling us that infinity exists as a physical object. Perhaps the theory is telling us that its language has reached its limit.

Infinite density is not an explanation.

It is the collapse of explanation.

Zero volume containing collapsed mass is not a satisfying physical reality. A place where laws break down should not be celebrated as a place in the universe. It may be the boundary where the old model loses coherence. UED avoids this problem by replacing infinite collapse with field saturation.

This is a more disciplined interpretation.

The darkness of a black-hole-like region does not have to be absence. Darkness can also arise from saturation. When a detector is overwhelmed, it may fail to give meaningful detail. When a signal is lost in noise, the message may not return, but that does not mean nothing exists. When a medium becomes too disturbed, a pattern may dissolve beyond recognition. The issue is not emptiness; the issue is recoverable coherence.

Under UED, the black-hole-like region is dark because the field no longer returns ordinary ripple information.

That is not a hole.

That is a saturation boundary.

The event horizon also changes meaning. It is no longer treated as a magical wall around a cosmic monster. It becomes the boundary of recoverable coherence. Beyond it, ordinary signals do not resolve back to the external observer. This does not prove an infinite-density center. It only tells us that the field condition has crossed a threshold beyond which normal observational recovery fails.

The event horizon marks the limit of readable field behavior.

Not necessarily the edge of reality.

This is why the monster myth dissolves.

A monster destroys. A Ripple Saturation Zone regulates.

A monster swallows. A saturation zone transforms coherence.

A monster is imagined as a thing. A saturation zone is understood as a condition.

This is especially important when we reinterpret galactic centers. Under the old image, the so-called supermassive black hole at the center of a galaxy becomes a destructive endpoint. It is a dark object sitting inside a luminous system. But under UED, the galactic center becomes a **Central Ripple Saturation Zone**, a regulatory core of the galactic field architecture.

It is not merely where matter dies.

It may be where the galaxy's field imbalance is managed.

It may concentrate gradients, absorb excess disturbance, anchor rotational coherence, and maintain structural balance. Its darkness is not necessarily destructive emptiness. It may be regulatory saturation. It may be the inward boundary that allows the outward luminous structure to remain coherent.

This changes the emotional identity of the black hole completely.

The dark center is no longer the villain of the galaxy.

It becomes the silent regulator.

It is not outside the galaxy's beauty. It belongs to the architecture of that beauty. The visible galaxy shines outward through ripple condensation, while the dark center holds inward through saturation. The two are not enemies. They are complementary expressions of one field system.

This also helps dissolve the problem of early supermassive black holes. In the standard model, they look anomalous because they appear too massive too soon. They must grow from seeds, feed rapidly, merge, accrete, and become enormous within a limited timeline. This creates a growth crisis.

But if the central region is a regulatory saturation zone born with the galactic field architecture, the crisis weakens. It does not need to be a separate object that grew like a cosmic animal. It may arise as the inward regulatory condition of the field event that also produces the galaxy outwardly.

The galaxy and its regulator may be one structure seen from two sides.

This is why black holes cease to be anomalies under UED.

The anomaly existed because the old interpretation forced extreme field behavior into the language of collapse, singularity, infinity, and monsterhood. It treated darkness as absence, boundary as wall, signal loss as destruction, and mathematical breakdown as physical truth.

UED reads the same phenomenon differently.

Darkness becomes saturation.

Boundary becomes coherence limit.

Extreme gravity becomes field-gradient intensity.

The galactic center becomes regulator.

The black hole becomes a field condition, not an impossible object.

This does not end all mystery. Ripple Saturation Zones still require deeper study. Their internal field behavior, interaction with surrounding structures, energy redistribution, and observational signatures must be developed further. But the mystery becomes cleaner. It no longer depends on impossible infinity. It no longer requires a cosmic monster. It no longer asks reality to break its own logic.

The black hole anomaly dissolves when darkness is read as saturation rather than absence or infinity.

The universe did not create monsters.

Our interpretation did.

12.4 — Redshift Without the Time-Machine Illusion

Redshift is real.

But the time-machine built around it may not be.

This distinction is essential. The signal shifts. The wavelength stretches toward red. Instruments measure it. Data records it. No serious reinterpretation needs to deny the observation. The problem begins when that observation is converted too quickly into a complete cosmic story.

Redshift becomes distance.

Distance becomes lookback time.

Lookback time becomes early universe.

Early universe becomes cosmic childhood.

Cosmic childhood becomes expected immaturity.

And then the universe is judged.

This chain has shaped modern cosmology so strongly that it now feels almost automatic. A high-redshift galaxy is not merely described as a structure whose signal is shifted. It is immediately placed into the early universe. Once placed there, it must behave according to the expectations of that placement. It should be young, faint, loose, forming, unfinished, and developmentally simple.

But JWST has disturbed this automatic confidence.

It has shown high-redshift structures that do not always behave like weak cosmic infants. Some appear bright. Some appear compact. Some appear structurally coherent. Some suggest central regulatory activity. Some appear mature enough to trouble the old timeline. The discomfort arises not merely because these structures exist, but because redshift has already assigned them a historical role.

The object is first placed into the past.

Then it is accused of being too mature for that past.

This is the time-machine illusion.

A telescope is treated as if it simply looks backward into old frames of a cosmic film. The farther it sees, the earlier the frame. The earlier the frame, the less mature the structure should be. Under this interpretation, JWST becomes a machine for viewing the childhood of the universe. Every high-redshift signal becomes a recovered scene from ancient cosmic history.

But Universal Energy Dynamics asks whether this is too simple.

A redshifted signal may not be merely a timestamp from a dead past. It may be a field-marked consequence arriving through Structured Vacuum Energy. It may carry delay, yes. But it may also carry strain, resistance, ripple tension, alignment delay, and the imprint of the field condition through which it has resolved.

In UED, redshift is not rejected.

It is widened.

It is no longer interpreted only as the stretching of light through expanding space. It is also read as a possible signature of field resistance. A signal does not travel through empty nothingness. It travels through condition. It emerges from a field state. It passes through layers of resistance and tension. It arrives as a transformed consequence.

This means redshift may not only tell us how far away something is.

It may tell us how deeply strained the field is.

A highly redshifted signal may indicate distance and delay, but it may also indicate that the signal has emerged through high field resistance. It may reflect compact energy configuration. It may reveal ripple tension. It may carry alignment delay caused by the state of Structured Vacuum Energy across the path of observation.

This changes the meaning of deep cosmic observation.

JWST may not be merely looking backward.

It may be looking deeper.

It may be resolving field layers that older instruments could not read. Its infrared sensitivity allows it to detect signals that were previously inaccessible. Those signals may belong to regions where the field condition differs strongly from our local observational state. When these signals are immediately turned into ancient history, the structures appear anomalous. But when they are read as field-depth signals, their intensity begins to make more sense.

A Little Red Dot, then, is not automatically a premature object from cosmic infancy.

It may be a compact field-condensation zone whose redshift carries resistance.

A bright high-redshift galaxy is not automatically a structure that grew too fast.

It may be a coherent field architecture whose signal has been strained through layered conditions.

A central regulator at cosmic depth is not automatically a black hole that became supermassive too soon.

It may be the inward saturation boundary of a galactic field system, misread through chronological expectation.

The old lens asks, "How can this exist so early?"

The UED lens asks, "What field condition does this signal reveal?"

This is the decisive shift.

Redshift without the time-machine illusion does not mean redshift loses value. It becomes more valuable. It stops being a single-purpose cosmic clock and becomes a layered diagnostic. It can still relate to distance. It can still involve delay. But it may also tell us about field depth, strain, resistance, and resolution behavior.

A number becomes richer when its meaning becomes layered.

The mistake is not measurement.

The mistake is reduction.

Redshift was reduced into distance. Distance was reduced into past. Past was reduced into earlyness. Earlyness was reduced into immaturity. Then the universe was called strange when it refused the reduction.

But perhaps the reduction was always too narrow.

The universe may not be sending us simple postcards from ancient time. It may be sending field consequences that become observable only after delay, resistance, and resolution. The telescope does not merely retrieve a dead photograph. It completes an observational relation in the present. The signal becomes data now. The interpretation happens now. The meaning is arranged now.

This does not erase history.

It disciplines history.

It reminds us that cosmic history is reconstructed from present signals. Those signals carry field conditions. If we ignore those conditions, history becomes rigid and misleading. If we include them, history becomes deeper and more honest.

This is why redshift without the time-machine illusion is necessary for the end of cosmic anomalies.

Many anomalies appear because redshift is used as a historical stamp before the physical nature of the signal is fully interpreted. Once the object is stamped as early, its maturity becomes a problem. But if redshift is first read as field strain, the same maturity may become meaningful. It may speak of coherence, condensation, saturation, and resistance.

The universe did not necessarily form impossible structures in its childhood.

We may have mistaken field depth for childhood.

We may have mistaken strain for age.

We may have mistaken delay for history.

We may have mistaken redshift for a time machine.

Under UED, redshift becomes less theatrical and more profound. It is not merely the universe saying, "This happened long ago." It may be the field saying, "This signal has carried resistance." It may be the visible color of tension inside Structured Vacuum Energy. It may be one of the clearest clues that cosmic observation is not a simple journey backward, but a layered reading of present-field consequence.

The time-machine illusion made the universe look impossible.

The field-strain interpretation makes it readable again.

12.5 — Hubble Tension Without Cosmic Crisis

The Hubble tension looks like a crisis only when the universe is expected to speak through one clean expansion rate.

That expectation is understandable. Modern cosmology has tried to describe the universe through a coherent expansion history. If the universe is expanding, then different methods of measuring that expansion should ideally agree. Local measurements should match early-universe-derived calculations. Nearby readings and deep-field inferences should converge into one stable cosmic value.

But they do not fully agree.

This disagreement has become one of the most important tensions in modern cosmology. It appears technical on the surface, but beneath the numbers lies a much deeper question: what exactly are we measuring when we say the universe is expanding?

The old interpretation assumes that expansion is a broad property of spacetime itself. Space expands. Light stretches. Redshift increases.

Distance and time are arranged into a cosmic history. From this view, disagreement in expansion rate becomes uncomfortable because one universe should not appear to expand at two different rates.

But Universal Energy Dynamics asks whether the discomfort arises from the interpretation, not the universe.

Perhaps the universe is not giving contradictory answers.

Perhaps different field states are resolving differently.

This is where Alignment Threshold Time becomes essential.

Under UED, time is not an independent cosmic road. Time is the measurable delay required for energy states to align, resolve, and become observable as consequence. If field conditions differ, alignment delays differ. If alignment delays differ, apparent rates differ. Therefore, what appears as a disagreement in expansion rate may actually be a disagreement in field-resolution condition.

This changes Hubble tension completely.

It is no longer simply a crisis in the expansion model. It becomes a diagnostic clue. It tells us that local measurements and deep-inferred measurements may not be reading the same type of field state. The local universe may represent a relatively stabilized field environment. Deep-field signals may carry stronger resistance, strain, ripple tension, and alignment delay. When both are forced into one smooth expansion narrative, tension appears.

But the tension may not be cosmic.

It may be interpretive.

The local reading may be valid within its field condition. The early-universe-derived reading may be valid within the assumptions and signal-state from which it is inferred. The disagreement may arise because the two readings are not simply two measurements of one uniform process. They may be two readings of different resolution states inside one structured field reality.

This is why UED does not panic before Hubble tension.

It listens to it.

A crisis becomes useful when it reveals a hidden assumption. Here, the hidden assumption is that the universe must be measured through one uniform expansion framework. But if the universe is Structured Vacuum

Energy, then rate cannot be separated from field condition. Redshift cannot be separated from strain. Distance cannot be separated from field depth. Time cannot be separated from alignment delay.

Once these are reunited, Hubble tension becomes less mysterious.

It may be the universe showing that apparent cosmic rate is state-dependent.

In the standard model, rate is treated as the expansion of space across time. In UED, rate may be the visible reading of field resolution across conditions. A stable field resolves one way. A strained field resolves another. A saturated field may resist ordinary coherence. A deep field layer may carry longer alignment delay. If we ignore these differences, we compress layered reality into one number.

Then the number refuses to settle.

This refusal is not failure.

It may be information.

A doctor does not treat differing symptoms as meaningless contradiction. Fever, pulse, pain, pressure, and inflammation may not all give the same type of reading, but together they reveal the condition of the body. Similarly, local expansion measurements, redshift data, deep-field signals, and background-inferred values may not be enemies. They may be different symptoms of the cosmic field.

The mistake is to demand that every symptom become one smooth value before understanding the underlying condition.

UED asks us to read the condition first.

What field state does the signal represent?

What resistance shaped it?

What alignment delay affected it?

What redshift strain is hidden inside it?

What resolution threshold made it observable?

These questions move Hubble tension from crisis to diagnosis.

The old question is: why do the expansion rates disagree?

The UED question is: what field states are hidden inside those rates?

The old question is: which measurement is wrong?

The UED question is: what delay is being mistaken for expansion?

The old question is: how can the universe have one history with two readings?

The UED question is: why did we assume one smooth history could contain all field states?

This is the deeper correction.

Hubble tension may be one of the strongest signs that the cosmic movie reel is too simple. A movie reel expects sequence. It expects the universe to unfold through one orderly timeline. It expects early and late, near and far, local and deep, all to fit into one continuous story. But a field reality may not behave like a film. It may resolve according to condition.

The local universe and the deep field may not simply be different scenes in one movie.

They may be different layers of field resolution.

This does not mean the universe is chaotic. It means it is structured. A structured system can contain different conditions without losing unity. The human body is one system, yet the brain, heart, lungs, muscles, blood, and skin do not share identical activity. A galaxy is one system, yet its center, arms, halo, and star-forming regions do not behave identically. Unity does not require sameness.

The universe may be one structured field, but its resolution states may differ.

Therefore, Hubble tension may not be the universe disagreeing with itself. It may be the universe refusing to be flattened into one uniform expansion narrative.

This is why the tension ceases to be an anomaly under UED. It is no longer a strange crack in reality. It becomes expected evidence that field condition matters. If redshift carries strain, if time is alignment delay, if distance includes field depth, and if expansion is energy redistribution through Structured Vacuum Energy, then different apparent rates are not shocking.

They are meaningful.

The crisis dissolves because the assumption changes.

The universe does not need to produce one clean rate for all observational conditions. It needs to produce consequences according to field state. If those consequences are later interpreted as expansion rate, the interpretation must include the state from which they emerged.

Hubble tension, then, is not an embarrassment.

It is a warning against oversimplification.

It tells us that cosmic measurement cannot be separated from the structure of the field being measured. It tells us that time cannot be treated as a universal road. It tells us that redshift cannot be reduced to distance alone. It tells us that expansion may not be a simple stretching of emptiness.

It may be field resolution becoming visible at scale.

The universe is not in crisis because our numbers do not immediately agree.

Our interpretation is in crisis because it expected layered reality to behave like a single smooth clock.

Under UED, Hubble tension becomes one more place where the anomaly ends. Not because the disagreement is ignored, but because the disagreement is given a deeper meaning. It is no longer the universe failing consistency. It is the field revealing state-dependent resolution.

The universe may not be expanding wrongly.

We may have been timing it too simply.

12.6 — The Unified UED Lens

Universal Energy Dynamics does not end cosmic anomalies by denying observation.

It ends them by changing the lens.

This distinction matters more than anything else in the final chapter. UED does not say galaxies are not observed. It does not say redshift is unreal. It does not say black-hole-like phenomena are imaginary. It does not say Hubble tension should be ignored. It does not say JWST data should be dismissed because it disturbs old models.

UED accepts the observation.

Then it questions the interpretation.

This is the mature path. The universe gives signals, structures, motions, darkness, redshift, and measurement tensions. The responsibility of theory is not to force these into inherited language, but to develop a frame large enough to read them coherently.

The older frame begins with objects.

UED begins with the field.

The older frame begins with matter.

UED begins with Structured Vacuum Energy.

The older frame begins with time as a cosmic road.

UED begins with consequence, alignment, resistance, and resolution.

This is why UED does not patch anomalies one by one. It does not say, "Here is one explanation for early galaxies, another for black holes, another for redshift, another for Hubble tension." That would become another patchwork. It would be no better than adding more exceptions to an already strained model.

Instead, UED changes the foundation from which all these problems arise.

Matter is not treated as dead substance floating in empty space. It is ripple condensation within Structured Vacuum Energy. This means galaxies are not merely piles of collected matter. They are field architectures.

Gravity is not treated as an invisible pull or merely the bending of empty spacetime. It is energy-gradient equilibrium. This means galactic motion, clustering, rotation, and structural coherence are not only the effects of mass pulling mass. They are expressions of field imbalance seeking balance.

Black holes are not treated as infinite-density monsters. They are Ripple Saturation Zones. This means darkness is not automatically absence, and signal loss is not automatically singularity. It may be coherence failure beyond observational recovery.

Galactic centers are not treated as accidental cosmic pits. They are central regulators. This means the dark core of a galaxy may not be a destructive endpoint, but the inward saturation boundary that helps maintain the outward luminous pattern.

Redshift is not treated only as a timestamp. It is also read as field strain, ripple tension, resistance, and alignment delay. This means high redshift does not automatically mean primitive childhood. It may indicate deeper field condition.

Time is not treated as an independent road. It is alignment delay and conscious narration. This means the universe may not be moving through a movie reel of past, present, and future. Field consequences resolve, and observers arrange them as time.

Cosmic depth is not treated only as pastness. It is also field depth. This means JWST may not merely be looking backward into ancient frames. It may be resolving deeper layers of structured present-field reality.

When these ideas are placed together, the anomalies begin to connect.

They are not separate accidents.

They are symptoms of one older lens reaching its limit.

The mature high-redshift galaxy, the Little Red Dot, the black hole, the event horizon, the central regulator, the redshift puzzle, the Hubble tension, and the failure of the cosmic movie reel all arise from the same interpretive mistake: treating a field universe as a matter-first, time-first, object-first story.

Once the universe is read as field consequence, the discomfort changes.

The galaxy is no longer too early. It is a coherent field architecture.

The Little Red Dot is no longer too compact. It is a high-tension condensation zone.

The black hole is no longer impossible. It is a saturation boundary.

The event horizon is no longer a magical wall. It is a limit of recoverable coherence.

The galactic center is no longer a monster. It is a regulator.

Redshift is no longer merely the color of distance. It is the color of resistance.

Hubble tension is no longer cosmic disagreement. It is state-dependent resolution delay.

Time is no longer the master road. It is the measurable shadow of consequence.

This is the unified UED lens.

It does not simplify the universe cheaply. It makes the universe coherent by reading its layers together.

The old framework separates everything. Matter here, space there. Gravity here, time there. Redshift here, distance there. Galaxy here, black hole there. Observation here, interpretation there. Then, after separating reality into pieces, it struggles to explain why the pieces do not fit cleanly.

UED reunites them.

Matter, gravity, redshift, time, structure, saturation, and observation are not separate mysteries. They are connected expressions of structured energy behavior.

This is why UED has explanatory strength. It does not merely offer alternative words. It offers a different grammar. In the old grammar, matter acts, gravity pulls, space expands, time carries, redshift timestamps, and anomalies disturb. In the UED grammar, the field conditions, energy condenses, gradients balance, saturation regulates, signals strain, time delays, and consequences become observable.

The same universe begins to speak differently.

This also explains why the end of anomalies does not mean the end of mystery.

The universe remains profound. Structured Vacuum Energy must still be explored. Ripple condensation must still be developed. Saturation zones must still be studied. Redshift as field strain must still be refined. Alignment Threshold Time must still be tested, modeled, and clarified. UED itself must grow through criticism, correction, and deeper formulation.

But the mystery becomes healthier.

It is no longer the mystery of contradiction. It becomes the mystery of depth.

Contradiction says, "This should not exist."

Depth says, "We have not yet understood the condition through which this exists."

That is a major difference.

Cosmic anomalies, under UED, do not disappear because the data is ignored. They disappear because the data is no longer forced into the wrong expectation. A galaxy is allowed to be coherent. A dark center is allowed to regulate. A redshifted signal is allowed to carry resistance. A measurement tension is allowed to reveal field-state difference. Time is allowed to become delay instead of destiny.

This is why the unified lens is necessary.

Without it, cosmology keeps repairing one crack at a time. A mature galaxy requires faster growth. A Little Red Dot requires special explanation. A black hole requires singularity management. A Hubble tension requires new parameters. A redshift puzzle requires recalibration. Every anomaly becomes its own emergency.

With UED, the emergencies become connected signals.

They all point toward one need: reinterpret the foundation.

The universe is not asking for endless patches.

It is asking for a deeper frame.

This is the final power of Universal Energy Dynamics. It does not fight observation. It listens to observation until the old language becomes insufficient. Then it gives new language: field, consequence, condensation, gradient, saturation, alignment, delay, coherence, resistance, and regulation.

Through this language, the cosmos becomes less broken.

Not because we force it into comfort.

But because we stop forcing it into the wrong discomfort.

The unified UED lens shows that the universe may be coherent at a deeper level than the old model could see. Its so-called anomalies may be field signatures. Its tensions may be diagnostic. Its darkness may be saturation. Its maturity may be coherence. Its redshift may be strain. Its time may be narration.

The end of cosmic anomalies begins here.

Not with the denial of mystery.

But with the end of misreading.

Closing Reflection — When the Frame Changes, the Cosmos Becomes Whole Again

The cosmos was never broken.

It was our frame that began to crack.

That is the final realization this book has been moving toward. The universe did not suddenly become strange because JWST looked deeper. Galaxies did not become impossible because they appeared brighter than expected. Little Red Dots did not become rebellious because they looked compact and intense. Black holes did not become monsters because mathematics reached infinity. Redshift did not become mysterious because signals shifted toward red. Hubble tension did not appear because the universe forgot its own rate.

These things became anomalies because the old frame could no longer hold them comfortably.

A frame is useful until reality outgrows it.

Then the same frame becomes a cage.

For a long time, modern cosmology gave us a powerful story. Matter filled space. Gravity organized matter. Spacetime expanded. Redshift measured distance. Distance gave lookback time. Lookback time gave cosmic history. The early universe was young. The later universe became mature. Galaxies grew slowly. Black holes formed through collapse. The universe moved like a movie reel from origin to present.

This story was elegant.

It was useful.

It was not meaningless.

But usefulness is not final truth.

When deeper observation began to reveal structures that resisted this timeline, the old story started producing discomfort. It did not collapse immediately. It stretched. It adjusted. It added possibilities. It searched for faster formation, heavier seeds, hidden scaffolds, special conditions, unusual efficiencies, and new parameters.

This is natural. Every mature science first tries to save its working model.

But when exceptions multiply, intelligence must ask whether the issue lies not in the universe, but in the interpretation.

Universal Energy Dynamics begins from that question.

It does not accuse the universe of contradiction. It does not dismiss observation. It does not say the telescope is wrong because the data is uncomfortable. It does something more disciplined: it changes the starting point.

Matter is not the beginning. It is ripple condensation within Structured Vacuum Energy.

Gravity is not merely pull or curved emptiness. It is energy-gradient equilibrium.

Black holes are not infinite-density monsters. They are Ripple Saturation Zones.

Galactic centers are not accidental pits. They are central regulators.

Redshift is not only a timestamp. It is also field strain and resistance.

Time is not a cosmic road. It is alignment delay and conscious narration.

Cosmic depth is not only pastness. It is field depth.

Once this lens is applied, the cosmos begins to reorganize before the mind.

Not the cosmos itself.

Our reading of it.

The mature galaxy is no longer accused of growing too fast. It is understood as a coherent field architecture. The Little Red Dot is no longer treated as an impossible infant of the early universe. It becomes a compact high-tension condensation zone. The black hole is no longer a supernatural pit. It becomes a saturation boundary where recoverable coherence fails. The event horizon is no longer a magical wall. It becomes an observational boundary. The galactic center is no longer a monster. It becomes the dark regulator of luminous structure. Redshift is no longer merely a time-machine stamp. It becomes the color of field resistance. Hubble tension is no longer cosmic confusion. It becomes state-dependent resolution delay.

This is how anomalies end.

Not by denying them.

Not by hiding them.

Not by forcing them into old comfort.

They end when the frame becomes deeper than the contradiction.

The end of cosmic anomalies does not mean the end of mystery. That would be a poor ending. Mystery must remain because reality is larger than any human theory, including UED. There will always be deeper measurements, sharper instruments, better questions, and more refined formulations. No serious thinker should claim that one book has closed the universe.

But there is a difference between mystery and contradiction.

Mystery invites exploration.

Contradiction exposes a broken frame.

The old cosmological frame turned too many observations into contradiction because it carried certain assumptions too far. It treated matter as primary. It treated vacuum as background. It treated time as road. It treated redshift as clock. It treated cosmic depth as pastness. It treated black holes as singularity-monsters. It treated Hubble tension as crisis.

UED does not eliminate mystery.

It removes unnecessary contradiction.

That is why the cosmos becomes whole again.

Wholeness does not mean simplicity. It does not mean every detail is easy. It does not mean the universe becomes small enough for human confidence. Wholeness means the pieces begin to belong together. Galaxies, redshift, gravity, time, darkness, motion, and tension stop appearing as isolated puzzles and begin appearing as connected expressions of field behavior.

The universe becomes coherent, not because we forced it to obey us, but because we finally stopped forcing it into the wrong story.

This is the humility required at the end.

The universe does not owe us agreement with our model.

Our model owes the universe deeper listening.

JWST has not embarrassed the cosmos. It has embarrassed overconfidence. It has not shown us that reality is confused. It has shown us that interpretation must evolve. Its deepest gift may not be only sharper images, but sharper discomfort. It has made old assumptions visible. It has forced cosmology to ask whether the universe is a movie reel or a field of consequences.

And once that question is honestly asked, the old certainty cannot return unchanged.

Perhaps we were never looking at a broken timeline.

Perhaps we were looking at a structured field.

Perhaps we were not seeing impossible galaxies.

Perhaps we were seeing coherent condensation.

Perhaps we were not seeing cosmic monsters.

Perhaps we were seeing saturation regulators.

Perhaps we were not looking backward into dead frames.

Perhaps we were looking deeper into layered reality.

Perhaps the universe was not strange.

Perhaps our interpretation was too narrow.

This is the final landing of the book.

The end of cosmic anomalies is not the end of science. It is the beginning of a cleaner science. A science that does not panic before contradiction. A science that does not worship old metaphors. A science that does not mistake mathematical breakdown for physical truth. A science that reads signals as consequences, not merely as confirmations of inherited models.

The cosmos was never broken.

The frame was.

And when the frame changes, the universe does not need to be repaired.

It only needs to be read again.

Then the galaxies return to coherence.

The darkness returns to function.

The redshift returns to depth.

The tension returns to meaning.

Time returns to consequence.

And the universe, no longer trapped inside our old interpretation, becomes whole again.

Closing Reflection

The Cosmos Was Never Broken

The cosmos was never broken.

It was our frame that began to crack.

That is the quiet realization this book has been moving toward from the beginning. The universe did not suddenly become strange because our instruments became sharper. Galaxies did not become impossible because JWST looked deeper. Little Red Dots did not become rebellious because they appeared compact, bright, and intense. Black holes did not become monsters because mathematics reached the edge of its own language. Redshift did not become mysterious because light shifted toward red. Hubble tension did not arise because the universe forgot how to expand.

These things became anomalies because our interpretation could no longer hold them comfortably.

For a long time, modern cosmology gave humanity a powerful story. Matter filled space. Gravity organized matter. Spacetime expanded. Redshift measured distance. Distance gave lookback time. Lookback time became cosmic history. The early universe was expected to be young, faint, immature, and incomplete. The later universe was expected to become structured, luminous, mature, and complex. The cosmos was imagined like a grand movie reel moving from origin to present.

This story was elegant.

It was useful.

It was not meaningless.

But usefulness is not final truth.

A map may guide a traveler for many miles, but when the terrain changes, the map must be checked again. A theory may organize enormous observation, but when sharper observation repeatedly creates discomfort, the theory must not be protected from questioning. Science becomes weak when it becomes more loyal to its inherited language than to the reality it is supposed to understand.

The universe has no obligation to obey our model.

Our model has the obligation to understand the universe.

That is the discipline this book has tried to preserve.

The aim was never to deny observation. The galaxies are there. The redshift is measured. The dark centers influence motion. The Hubble tension remains meaningful. JWST has shown structures that deserve serious attention. None of this should be dismissed. But the meaning of these observations depends on the frame through which they are read.

If the frame is matter-first, time-first, and movie-reel-based, then many observations become uncomfortable.

A mature galaxy becomes too early.

A compact red structure becomes too intense.

A galactic center becomes an impossible monster.

A high-redshift signal becomes an ancient frame.

A disagreement in expansion readings becomes crisis.

But when the frame changes, the same observations begin to speak differently.

Matter becomes ripple condensation within Structured Vacuum Energy. Gravity becomes energy-gradient equilibrium. Black holes become Ripple Saturation Zones. Galactic centers become regulatory cores. Redshift becomes field strain, ripple tension, and resistance. Time becomes Alignment Threshold Time and conscious narration. Cosmic depth becomes field depth, not merely pastness.

Then the anomaly begins to dissolve.

Not because the mystery is removed.

Because the contradiction weakens.

This is the difference between mystery and confusion. Mystery belongs to reality. Confusion often belongs to interpretation. Mystery invites deeper intelligence. Confusion forces us to repair wrong assumptions again and again. The universe will always remain mysterious because reality is larger than any framework. But many contradictions disappear when the lens becomes deep enough to read what was previously misread.

The end of cosmic anomalies does not mean the end of science.

It means the beginning of cleaner science.

A science that does not panic before discomfort. A science that does not worship old metaphors. A science that does not mistake mathematical breakdown for physical truth. A science that does not turn every difficult signal into a patching exercise. A science that asks, with humility and courage, whether the problem is in the cosmos or in the frame.

Universal Energy Dynamics offers such a frame.

It does not claim that every detail is finished. It does not ask the reader to abandon critical thinking. It does not present itself as a decorative replacement for established science. It asks for a deeper reading: field before object, consequence before chronology, alignment before time, saturation before singularity, resistance before redshift-as-clock, coherence before maturity-as-age.

Through this lens, the universe becomes whole again.

The galaxy is not premature.

It is coherent.

The black hole is not a monster.

It is saturation.

The redshift is not merely a timestamp.

It is resistance made visible.

The Hubble tension is not cosmic confusion.

It is state-dependent resolution delay.

Time is not the road beneath the universe.

It is the shadow cast by consequences as they become observable.

This is not the end of wonder. It is the end of blaming the cosmos for the limits of our reading. The universe was never late, premature, confused, or broken. These were words born from comparison with a script the universe never agreed to follow.

Perhaps there was no script.

Perhaps there was only reality, and human intelligence kept writing stories around it.

Now, sharper observation has begun to challenge those stories. That challenge should not frighten us. It should mature us. Because every true expansion of knowledge begins when reality refuses our comfort.

The cosmos was never broken.

The frame was.

And when the frame changes, the universe does not need to be repaired.

It only needs to be read again.

Then the galaxies return to coherence.

The darkness returns to function.

The redshift returns to depth.

The tension returns to meaning.

Time returns to consequence.

And the universe, no longer trapped inside our old interpretation, becomes whole again.

Structural Summary

The End of Cosmic Anomalies is built around a single structural claim: the universe may not be producing impossible realities; modern cosmology may be interpreting reality through a frame that has become too narrow.

The book begins by questioning the repeated use of the word "anomaly" in modern cosmology. Mature galaxies at cosmic depth, compact Little Red Dots, black-hole-like regions, redshift puzzles, and the Hubble tension are usually treated as problems inside the cosmic timeline. This book does not dismiss those observations. It takes them seriously. But it asks whether the discomfort arises from the universe itself or from the assumptions used to interpret it.

The central argument is that many cosmic anomalies are created at the meeting point between observation and expectation. A galaxy is not "too early" by itself. It becomes too early only when placed inside a timeline that expects cosmic infancy at depth. A black hole is not automatically a monster. It becomes one when interpreted through collapse, singularity, and infinite density. Redshift is not automatically a complete historical clock. It becomes one when distance is converted too quickly into pastness. Hubble tension is not automatically a cosmic crisis. It becomes one when the universe is expected to speak through one uniform expansion rate.

The book first exposes this crisis of interpretation. It shows how modern cosmology has built a powerful but restrictive cosmic movie reel: beginning, expansion, cooling, particles, stars, galaxies, maturity, and present observation. Within this model, deeper space is treated as earlier time, and earlier time is expected to contain immature structures. The James Webb Space Telescope becomes important because it disturbs this expectation. It does not break the universe; it reveals the stress in the old interpretive frame.

From there, the book introduces Universal Energy Dynamics as a deeper lens. UED does not begin with dead matter, empty space, and time as an independent cosmic road. It begins with structured energy, field condition, ripple behavior, resistance, alignment, saturation, and consequence. This shift changes the meaning of almost every major cosmological concept discussed in the book.

Matter is reinterpreted as ripple condensation within Structured Vacuum Energy, not as an independent substance floating in empty space. Gravity becomes energy-gradient equilibrium, not merely invisible pull or the bending of empty spacetime. Black-hole-like regions become Ripple Saturation Zones, not infinite-density monsters. Galactic centers become regulatory cores, not merely destructive endpoints. Redshift becomes a layered signal carrying field strain, ripple tension, resistance, and alignment delay, not only a timestamp of cosmic past.

This field-first interpretation changes how cosmic maturity is understood. A mature-looking high-redshift galaxy may not be a chronological contradiction. It may be a coherent field architecture. A Little Red Dot may not be an impossible early object. It may be a compact high-tension condensation zone. A central black-hole-like region may not have grown too fast. It may have arisen with the galaxy as part of one field-realignment event.

The book then moves into its deepest conceptual shift: the reinterpretation of time. It argues that time should not be treated as an independent cosmic road on which the universe travels. Time is real as experience, measurement, duration, and sequence, but it may not be fundamental in the way modern cosmology often assumes. Under UED, time becomes Alignment Threshold Time: the measurable delay required for energy states to align, resolve, and become observable as consequence.

This idea reframes the Hubble tension. If different field states resolve through different alignment delays, then different expansion-rate readings may not be contradictions. They may be evidence of state-dependent resolution delay. The universe may not be expanding wrongly; we may be timing it too simply.

The book also challenges the assumption that telescopes are simply time machines looking backward into dead cosmic frames. Signal delay is real, but delay alone does not prove that every distant observation is a direct view of primitive cosmic childhood. Cosmic depth may also be field depth. JWST may not merely be looking backward; it may be resolving deeper layers of present structured field reality.

Finally, the book brings the entire argument together by showing that the anomalies are not isolated. Galaxies, black holes, redshift, cosmic maturity, time, and Hubble tension are connected through one interpretive shift. When the universe is read as a matter-first, time-first, movie-reel system, many observations become strange. When it is read as a structured field of consequences, those same observations begin to form a coherent pattern.

The book does not claim that all mystery ends. It does not claim that every detail is settled. It does not reject science, observation, or measurement. Instead, it argues that mystery must be separated from confusion. Mystery belongs to reality. Confusion often belongs to interpretation.

The structural movement of the book is therefore clear: it begins with anomalies, questions the frame that produces them, introduces UED as a field-based reinterpretation, dismantles the cosmic movie reel, and ends with the claim that the cosmos was never broken.

The frame was.

When the frame changes, galaxies return to coherence, darkness returns to function, redshift returns to depth, Hubble tension returns to meaning, and time returns to consequence.

Closing Note

When I began thinking about this book, I was not trying to create another disagreement with mainstream science.

There are already enough disagreements in the world. Every field has its groups, schools, loyalties, and guarded positions. Science too, though it tries to remain objective, is not free from this human tendency. Once a model becomes successful, people begin to protect it. Once a language becomes dominant, it becomes difficult to think outside that language.

But reality does not care about our comfort.

That is where this book begins and ends.

For me, *The End of Cosmic Anomalies* is not only a book about galaxies, black holes, redshift, Hubble tension, or JWST. These are important, no doubt. But the deeper subject is interpretation. The real question is not only what we observe. The real question is how we read what we observe.

Again and again, I have felt that the universe is not confused. The cosmos is not producing strange things to disturb us. It is we who become disturbed when reality does not obey the story we have already accepted. A galaxy does not become "too early" by itself. A black hole does not become a monster by itself. Redshift does not become a complete history by itself. Hubble tension does not become a crisis by itself.

These meanings are added by interpretation.

That is why I wanted to write this book in a direct but patient way.

I did not want to attack science. I wanted to question the frame through which science is reading the cosmos. There is a difference. To attack science is immature. To question interpretation is necessary. Science survives not because it protects old models, but because it keeps the courage to revise them.

Universal Energy Dynamics is my attempt to offer such a revision.

UED begins with a simple but powerful shift: do not start with dead matter, empty space, and time as a cosmic road. Start with structured energy, field condition, consequence, resistance, alignment, saturation, and equilibrium-seeking behavior. Once this starting point changes, many so-called anomalies begin to look different.

Matter becomes ripple condensation.

Gravity becomes energy-gradient equilibrium.

Black holes become Ripple Saturation Zones.

Galactic centers become central regulators.

Redshift becomes field strain and resistance.

Time becomes Alignment Threshold Time and conscious narration.

Cosmic depth becomes field depth, not merely pastness.

This does not mean all questions are finished. In fact, the opposite is true. A new framework should not close inquiry; it should open better inquiry. If UED is to grow, it must face criticism, refinement, mathematical development, observational comparison, and deeper scientific discipline. I fully accept that. No serious idea should fear testing.

But before testing can become meaningful, a new question must first be allowed.

That is what this book tries to do.

It asks whether the universe is really producing anomalies, or whether our model is producing anomaly categories. It asks whether JWST has broken cosmology, or whether it has simply shown that older assumptions are no longer enough. It asks whether the universe is truly a historical movie reel, or whether we are looking into layered field reality through delayed and strained signals.

These questions may sound bold.

But every honest question sounds bold before it becomes necessary.

I have written this book as an independent researcher, not from the safety of a large institution, not from the shelter of academic authority, and not from the desire to decorate known ideas with new words. I have written it because I genuinely feel that the universe must be read through a deeper lens. The current language has carried science far, but perhaps it cannot carry us all the way.

At some point, intelligence must ask: what if the frame itself is tired?

This book is my answer to that moment.

I know very well that many readers may disagree. That is fine. Disagreement is not a problem. Blind acceptance is more dangerous than disagreement. I only request the reader to disagree honestly. Do not reject the idea merely because it is outside familiar terminology. Also, do not accept it merely because it sounds new. Read it, challenge it, test it, and then think.

That is all I ask.

Because the purpose of knowledge is not to win an argument.

The purpose of knowledge is to make reality clearer.

If this book helps even a few readers look at the cosmos differently, if it helps them separate observation from interpretation, if it helps them question the easy use of the word "anomaly," then the book has done its work.

The cosmos was never broken.

Perhaps our reading was.

And if our reading was broken, then there is no shame in repairing it. In fact, that is the dignity of intelligence. The greatest loyalty is not loyalty to an old model. The greatest loyalty is loyalty to truth, even when truth asks us to stand outside comfort.

I do not see this book as the end of a discussion.

I see it as an opening.

A new lens.

A new grammar.

A new invitation to read the universe again.

Not as a broken machine.

Not as a movie reel of dead history.

Not as a warehouse of impossible objects.

But as a structured field of consequences, still deeper than our language, still larger than our models, still waiting for intelligence to mature enough to read it with humility.

The universe does not need our permission to be what it is.

We only need the courage to see it again.

Core Concepts

This glossary is not meant to reduce the book into rigid definitions. It is meant to help the reader hold the central language of *The End of Cosmic Anomalies* with clarity.

The terms below belong to the interpretive framework of Universal Energy Dynamics. Some terms challenge mainstream cosmological vocabulary. Some reinterpret familiar scientific concepts. Some are newly framed within the UED lens.

They should be read not as decorative terminology, but as working concepts through which the book reopens the meaning of galaxies, black holes, redshift, cosmic depth, time, and so-called anomalies.

Anomaly

An anomaly is not simply something strange. It is something that does not fit the model through which it is interpreted.

Reality does not produce anomaly labels. Models do. A galaxy, black-hole-like region, redshifted signal, or expansion-rate disagreement becomes anomalous only when it refuses the expectations of a framework.

In this book, many cosmic anomalies are treated as signs that the interpretive frame has become too narrow, not as evidence that the universe itself is confused.

Alignment

Alignment refers to the process by which energy states, field conditions, or structural relationships reach a condition of coherence, stability, or observable consequence.

In UED, alignment does not mean moral harmony or human agreement. It means a field condition reaching the threshold where a consequence becomes readable, stable, or structurally expressed.

A galaxy may appear when field conditions align. A signal may become observable when ripple behavior resolves. A system may stabilize when gradients enter a balanced relationship.

Alignment Delay

Alignment delay is the duration required for a field condition or energy state to resolve into observable consequence.

This delay is not empty waiting. It is the active process through which energy interacts, resists, adjusts, aligns, saturates, or transforms.

Alignment delay is central to the reinterpretation of time, redshift, and Hubble tension in this book.

Alignment Threshold Time / ATT

Alignment Threshold Time is the measurable delay required for energy states to rearrange, align, and become observable as consequence.

ATT does not treat time as an independent cosmic road. It treats time as the duration of resolution between condition and consequence.

Different field states may carry different alignment thresholds. A stable field, strained field, compact field, or saturated field may not resolve in the same way. This is why ATT becomes important in interpreting Hubble tension and cosmic timing.

Black Hole

In mainstream imagination, a black hole is often treated as a cosmic monster: a collapsed object with an event horizon, infinite-density singularity, and no-return boundary.

This book does not deny black-hole-like phenomena. It questions the inherited interpretation.

Under UED, the so-called black hole is better understood as a Ripple Saturation Zone, a field condition where ordinary coherence fails to return to external observation.

Central Regulator

A central regulator is the dark regulatory core of a galactic field architecture.

What mainstream cosmology calls a supermassive black hole at the center of a galaxy is reinterpreted in this book as a Central Ripple Saturation Zone. It may not be merely a destructive endpoint, but a regulatory condition that helps maintain galactic coherence.

The central regulator holds inward saturation while the galaxy expresses outward luminosity.

Central Ripple Saturation Zone

A Central Ripple Saturation Zone is a Ripple Saturation Zone located at the regulatory core of a galaxy.

It is not treated as a singularity or infinite-density object. It is understood as a field saturation condition where ripple coherence becomes unrecoverable and gradient intensity becomes concentrated.

Its function may be regulatory rather than merely destructive. It may anchor motion, manage field imbalance, and help maintain large-scale galactic structure.

Coherence

Coherence means organized, stable, or readable field behavior.

A structure is coherent when its field condition is stable enough to appear as form, pattern, motion, or measurable identity. Coherence does not necessarily mean chronological age.

This is important for mature high-redshift galaxies. Under UED, a galaxy may appear mature because it is coherent, not necessarily because it is old in the conventional timeline.

Cosmic Depth

Cosmic depth is usually treated as distance into the past. In this book, cosmic depth is widened.

It may include spatial separation and signal delay, but it may also include field depth: layered resistance, strain, density, ripple tension, and alignment delay within Structured Vacuum Energy.

Thus, looking deeper may not always mean looking backward in a simple historical sense. It may mean resolving deeper layers of field condition.

Cosmic Movie Reel

The cosmic movie reel is the inherited image of the universe as a sequence of frames moving from beginning to present.

In this view, distant light becomes an earlier frame, high redshift becomes cosmic childhood, and deep observation becomes a form of time travel. This book argues that such a metaphor may be too simplistic.

UED replaces the movie reel with a field-based view: the universe is not a film strip of dead frames, but a structured field of consequences becoming observable through alignment, resistance, and resolution.

Cosmic Maturity

Cosmic maturity usually means developmental age within the standard cosmic timeline.

In UED, maturity can also mean field coherence. A galaxy may appear mature because its field architecture is organized, compact, luminous, and self-sustaining, not merely because it has existed for a long chronological duration.

This distinction is essential for interpreting JWST's mature-looking high-redshift galaxies.

Dark Matter

Dark matter is the unseen mass inferred in mainstream cosmology to explain gravitational behavior, galactic rotation, clustering, and large-scale structure.

This book does not casually dismiss dark matter inference, but it asks whether some gravitational behavior attributed to unseen matter may instead reflect deeper field-gradient behavior within Structured Vacuum Energy.

Under UED, the "missing" factor may not always be hidden material. It may be misunderstood field behavior.

Dark Energy

Dark energy is the concept used in mainstream cosmology to explain the apparent accelerated expansion of the universe.

In this book, dark energy is not treated as final explanation. UED asks whether what is interpreted as acceleration may instead be connected to field redistribution, alignment delay, vacuum gradient behavior, and state-dependent resolution within Structured Vacuum Energy.

Distance

Distance is usually understood as spatial separation.

UED does not deny spatial distance, but it adds another layer: field depth. A signal may be far in space, delayed in arrival, and also shaped by resistance, strain, and alignment conditions of the field.

This is why distance cannot automatically be reduced to pastness.

Energy-Gradient Equilibrium

Energy-gradient equilibrium is the UED reinterpretation of gravity.

Where energy density differs, the field develops gradient behavior. That gradient produces directional tendency as the system seeks balance. What we call gravity may be the observable consequence of such imbalance seeking equilibrium.

This shifts gravity from invisible pull or empty spacetime curvature to field-resolution behavior.

Event Horizon

In mainstream language, the event horizon is the boundary beyond which nothing can return from a black hole.

In UED, the event horizon is reinterpreted as an observational coherence boundary. It marks the limit beyond which ordinary ripple information cannot return in a recoverable form to the external observer.

It is not necessarily a magical wall or proof of a singularity. It may be the boundary of readable field behavior.

Field

A field is the underlying condition within which energy behavior, matter condensation, motion, signal propagation, and structural coherence occur.

In this book, the field is not treated as empty background. It is the active medium of reality. Under UED, the universe is fundamentally field-based rather than object-based.

Field Architecture

Field architecture refers to an organized large-scale field structure.

A galaxy, under UED, is not merely a pile of stars, gas, dust, and matter. It is a field architecture expressed through ripple condensation, rotational coherence, energy gradients, central regulation, and structural balance.

This concept is central to the reinterpretation of galaxies and their central regulators.

Field Coherence

Field coherence is the condition in which field behavior remains organized enough to produce stable structure, signal, identity, or measurable consequence.

Where coherence is sustained, matter, galaxies, stars, and signals can become readable. Where coherence fails, darkness, saturation, or signal loss may appear.

A Ripple Saturation Zone is where coherence becomes unrecoverable to external observation.

Field Condition

Field condition refers to the state of Structured Vacuum Energy in a given region or relation.

A field condition may be stable, strained, resistant, compact, dense, aligned, saturated, or unresolved. Under UED, the observed behavior of structures and signals depends not only on objects, but on the field conditions through which they arise and are measured.

Field Depth

Field depth means layered difference in field condition.

It includes resistance, strain, energy density, ripple tension, compactness, saturation, and alignment delay. Field depth is not identical to spatial distance or historical past.

A telescope may be resolving field depth, not merely looking backward through time.

Field Realignment

Field realignment is the process by which a field condition reorganizes into a new coherent state.

Galaxies may arise through field realignment. Central regulators may appear as part of the same realignment event. Redshifted signals may carry traces of alignment delay.

Field realignment is central to the UED explanation of rapid or mature-looking cosmic structure.

Field Resistance

Field resistance is the resistance encountered by energy, ripple behavior, or signals as they emerge, propagate, align, or resolve through Structured Vacuum Energy.

Redshift, delay, tension, distortion, or signal loss may carry signs of resistance. In UED, resistance is not merely obstruction; it is part of how the field produces measurable consequence.

Field Strain

Field strain is the stress or tension within a field condition caused by energy differences, density, resistance, ripple tension, or unresolved gradients.

In this book, redshift is partly reinterpreted as a possible signature of field strain. A redshifted signal may not merely indicate distance. It may also carry evidence of strained field interaction.

Galaxy

A galaxy is reinterpreted under UED as a field architecture, not merely a collection of stars.

It is an organized field consequence expressed through matter condensation, luminosity, rotation, central regulation, and energy-gradient behavior.

The galaxy does not simply gather slowly from matter. It may arise when field conditions reach coherence.

Gravity

Gravity is traditionally understood as attraction between masses or as the curvature of spacetime.

In UED, gravity is reinterpreted as energy-gradient equilibrium. It is the visible behavior of imbalance seeking balance within Structured Vacuum Energy.

Gravity does not "pull" in the crude sense. It reveals where field balance is unfinished.

Hubble Tension

Hubble tension is the disagreement between different methods of estimating the expansion rate of the universe.

In mainstream cosmology, it is often treated as a crisis. Under UED, it may indicate state-dependent resolution delay. Different field states may produce different alignment delays, and those delays may be misread as conflicting expansion rates.

Hubble tension may not be the universe disagreeing with itself. It may be the universe revealing that rate depends on field condition.

Interpretation

Interpretation is the framework through which observation is given meaning.

This book repeatedly separates observation from interpretation. A signal, structure, or measurement may be real, while the meaning assigned to it may be incomplete.

The central claim of the book is that many cosmic anomalies arise not from observation itself, but from interpretation.

JWST / James Webb Space Telescope

JWST is the observational trigger of this book.

It has revealed deep-field structures that challenge expectations of the standard cosmic timeline. In this book, JWST is treated not merely as a time machine looking backward into the early universe, but as an instrument capable of resolving deeper field layers.

JWST did not break the universe. It exposed the limits of an older interpretive frame.

Little Red Dots

Little Red Dots are compact red objects discussed in relation to JWST deep-field observations.

In mainstream interpretation, they may appear puzzling because of their compactness, brightness, redness, and apparent early placement.

Under UED, they may be read as compact high-tension ripple-condensation zones, where field strain, resistance, and energy density become visible.

Local Motion

Local motion refers to movement within the universe: planets orbiting, stars moving, galaxies rotating, signals propagating, and structures changing.

UED does not deny local motion. It distinguishes local motion from total cosmic travel. Things move within the universe, but the universe as totality may not move as an object through an external road.

Matter

Matter is traditionally treated as physical substance.

In UED, matter is ripple condensation within Structured Vacuum Energy. It is not the first truth, but a field consequence. Matter becomes visible when structured energy reaches stable coherence.

This reinterpretation is foundational for the entire book.

Model

A model is a disciplined framework for interpreting reality.

Models are necessary, but they are not reality itself. An anomaly often appears when reality refuses the expectations of a model. This book argues that the standard matter-first, time-first model may have become too narrow for newer observations.

Motion

Motion is the observed change of position, relation, or state.

Under UED, motion is not merely object travel through empty space. It is the visible consequence of field imbalance resolving through gradients, resistance, and alignment.

Motion is real locally, but it does not prove that the universe as totality travels through time.

Past

The past is the mind's arrangement of resolved consequences.

This does not make the past meaningless. It means the past is not necessarily an independent cosmic region stored behind us. In cosmology, what we call the past may be reconstructed from delayed signals, traces, and field-marked consequences received in the present.

Present

The present is the active interface where consequence becomes observable.

In human experience, the present is where attention meets reality. In cosmology, observation occurs in the present when a signal reaches an instrument and becomes measurable.

The present is not merely a clock moment. It is a resolution point.

Redshift

Redshift is the observed shift of light toward longer wavelengths.

Mainstream cosmology often interprets redshift as evidence of expansion, distance, and lookback time. UED widens the meaning: redshift may also carry field strain, ripple tension, resistance, and alignment delay.

Redshift may not be only the universe's timestamp. It may be the color of resistance.

Ripple

A ripple is a disturbance or propagating pattern within Structured Vacuum Energy.

In UED, ripple behavior helps explain matter condensation, signal propagation, field interaction, and saturation. Matter itself is treated as stabilized ripple condensation.

Ripple Condensation

Ripple condensation is the process through which structured energy stabilizes into matter-form.

Matter is not independent substance. It is condensed ripple behavior within Structured Vacuum Energy. This concept replaces the idea of dead material objects floating inside empty space.

Ripple Saturation Zone

A Ripple Saturation Zone is a region where field coherence collapses beyond ordinary observational recovery.

It is UED's reinterpretation of a black-hole-like region. It is not a hole, singularity, or infinite-density object. It is a field condition where ripple information no longer returns in a coherent form to the external observer.

Ripple Tension

Ripple tension refers to strain, disturbance, or pressure within ripple behavior as it propagates, aligns, resists, or saturates through the field.

Redshifted signals may carry ripple tension. Compact structures may express intense ripple tension. Saturation zones may form when ripple tension crosses coherence limits.

Signal

A signal is a measurable field consequence reaching observation.

A signal is not merely a message from the past. It may carry delay, resistance, strain, alignment history, and field-condition imprint.

In this book, signals are read as field consequences rather than simple historical photographs.

Signal Delay

Signal delay is the time required for a signal to reach observation.

UED does not deny signal delay. It questions whether delay alone justifies turning every distant observation into a complete historical frame. Signal delay is physics; cosmic history is interpretation built upon it.

Singularity

A singularity is traditionally described as a point of infinite density where physical laws break down.

In this book, singularity is treated not as final physical reality, but as a warning sign that a model has reached its interpretive limit.

UED replaces singularity with saturation.

Spacetime Fabric

Spacetime fabric is the popular metaphor used to explain gravity as curvature.

This book argues that the metaphor may have become too literal. If spacetime is empty, what bends? If it behaves, can it be nothing? UED replaces empty spacetime fabric with Structured Vacuum Energy as the active field condition.

State-Dependent Resolution Delay

State-dependent resolution delay means that different field states resolve into observable consequence at different rates.

A stable field and a strained field may not align the same way. A saturated region may fail to return coherence. This concept is central to the UED interpretation of Hubble tension.

Structured Vacuum Energy / SVE

Structured Vacuum Energy is the foundational medium in UED.

It is not empty space. It is the structured field condition from which matter, gravity, ripple behavior, redshift, saturation, and observable consequences arise.

SVE replaces the idea of dead vacuum with an active field reality.

Time

Time is traditionally treated as an independent dimension or cosmic road.

Under UED, time is causal sequencing, alignment delay, and conscious narration. It is not denied, but repositioned. Time is the measurable shadow of consequences as they become observable.

Time does not create causality. Causality creates the experience and measurement of time.

Total Stillness

Total stillness refers to the idea that the universe as totality may not move through anything outside itself.

Local motion exists within the universe, but the universe as a whole has no external road, background, or container through which it travels. It may be still as totality while active through consequences.

UED / Universal Energy Dynamics

Universal Energy Dynamics is the interpretive framework used throughout this book.

It begins with the idea that existence is a continuous consequence of structured energy, equilibrium-seeking behavior, ripple interaction, field resistance, alignment, and saturation.

UED reinterprets matter, gravity, black holes, redshift, time, cosmic maturity, and Hubble tension through a field-first lens.

Vacuum

In standard imagination, vacuum often means empty space.

In UED, vacuum is not absence. It is Structured Vacuum Energy. It is the active field condition through which observable reality manifests.

A behaving vacuum cannot be pure nothingness.

Whole / Wholeness

Wholeness means that the universe is not a broken collection of disconnected anomalies.

When read through UED, galaxies, gravity, redshift, black holes, time, and Hubble tension become connected expressions of field behavior.

Wholeness does not mean simplicity. It means coherence across layers.

Final Note on the Glossary

The purpose of this glossary is not to close discussion. It is to open the reader's understanding of the book's internal language.

The central shift is simple:

The universe is not primarily a collection of dead objects moving through empty space and time.

It is a structured field of consequences.

Once this is understood, many so-called anomalies stop appearing as cosmic failures and begin appearing as signals of deeper reality.

Key Insights

The following key insights capture the central arguments, conceptual turns, and quotable conclusions of *The End of Cosmic Anomalies*. They may be used by readers as a quick reference to the book's intellectual spine.

They are not substitutes for the full argument. They are distilled reminders of the lens through which the book reinterprets galaxies, black holes, redshift, time, Hubble tension, and the meaning of cosmic anomaly itself.

1. The universe did not become strange overnight.

The cosmos did not suddenly begin producing contradictions. What changed was our observational depth. As instruments became sharper, old assumptions became more exposed. The universe was not becoming stranger; it was becoming harder to misread.

2. An anomaly is often a failure of the frame, not a failure of reality.

An object becomes anomalous only when it does not fit the expectation imposed upon it. A mature galaxy, redshifted signal, black-hole-like region, or Hubble tension is not strange by itself. It becomes strange when the model cannot read it properly.

3. The problem is not observation; the problem is interpretation.

The book does not deny data. It questions the meaning assigned to data. JWST observations, redshift measurements, dark galactic centers, and Hubble tension must be taken seriously, but not automatically forced into inherited interpretations.

4. JWST did not break the universe; it broke the comfort of the timeline.

JWST revealed structures that disturbed the expected cosmic sequence. It did not create crisis. It exposed the hidden stress already present in the matter-first, time-first model of cosmology.

5. Cosmic depth may not always mean cosmic past.

Deep observation may involve signal delay, but delay alone does not justify turning every distant signal into a simple ancient frame. Cosmic depth may also mean field depth: resistance, strain, ripple tension, compactness, and alignment delay.

6. Looking deeper may not always mean looking backward.

A telescope may not be merely a time machine. It may also be a field-depth instrument resolving deeper layers of Structured Vacuum Energy. What appears as old light may also be a field-marked consequence arriving through delay.

7. Redshift may not be only a timestamp.

Redshift is real, but its meaning may be layered. It may indicate distance and delay, but it may also carry field strain, ripple tension, resistance, compact energy configuration, and alignment delay.

8. Redshift may be the color of resistance.

The red shift of a signal may not only tell us that the source is far away. It may reveal that the signal has travelled through or emerged from a strained field condition. Under UED, redshift becomes a diagnostic of field behavior.

9. Matter is not a dead thing; it is a field consequence.

Matter is not treated as independent substance floating inside empty space. It is ripple condensation within Structured Vacuum Energy. What appears as material form is stabilized field behavior.

10. Mass is not the final cause; it is a field-density signature.

Mass should not be worshipped as the deepest source of gravity. It may be the measurable signature of energy condensation within the field. Where the field condenses, mass becomes observable.

11. Gravity does not pull; it reveals imbalance.

Gravity is reinterpreted as energy-gradient equilibrium. It is not merely invisible attraction or bending of empty spacetime. It is the observable behavior of field imbalance seeking balance.

12. Empty spacetime cannot behave like something while being treated as nothing.

If spacetime bends, ripples, guides motion, and responds to matter, then it cannot be pure emptiness in any meaningful physical sense. UED replaces empty spacetime fabric with Structured Vacuum Energy as the active field condition.

13. Galaxies are not piles; they are field architectures.

A galaxy is not merely a collection of stars, gas, dust, and matter. It is a large-scale field architecture expressed through ripple condensation, luminosity, rotation, gradient behavior, and central regulation.

14. Cosmic maturity may mean coherence, not chronological age.

A galaxy may appear mature because its field architecture is coherent, not necessarily because it is old within a conventional timeline. This reframes mature high-redshift galaxies and Little Red Dots.

15. Little Red Dots may be high-tension field condensation zones.

Compact red structures observed at cosmic depth may not be impossible early objects. They may be concentrated expressions of field strain, ripple tension, resistance, and compact energy condensation.

16. The slow assembly model may be a trap.

If galaxies are assumed to form only through slow mechanical accumulation, then mature-looking deep structures become anomalies. If galaxies are field consequences, coherence may appear when field conditions reach threshold.

17. Black holes may not exist as we imagine them.

The observed phenomena are real, but the inherited image of a cosmic monster, infinite density, and physical hole may be interpretive overreach. UED reframes black-hole-like regions as Ripple Saturation Zones.

18. A singularity may be a warning sign, not a place.

When mathematics produces infinite density and law-breakdown, it may not be revealing nature's final object. It may be revealing the limit of the model's physical meaning.

19. Darkness is not always emptiness.

What appears dark may be saturated rather than absent. A black-hole-like region may be a field condition where ordinary ripple coherence fails to return to the external observer.

20. The event horizon may be a coherence boundary.

The event horizon is reinterpreted as a boundary of recoverable signal coherence, not a magical wall around an infinite-density object. It marks the limit of readable field behavior.

21. Galactic centers may be regulators, not monsters.

The dark center of a galaxy may not be a destructive endpoint. It may be a Central Ripple Saturation Zone that helps maintain galactic coherence, absorb field disturbance, and anchor large-scale structure.

22. The galaxy and its central regulator may arise together.

The chicken-and-egg problem of galaxies and supermassive black holes may be created by treating one field architecture as two separate objects. Under UED, outward galaxy structure and inward central regulation may arise as one field-realignment event.

23. The universe may not be a movie reel.

The cosmic movie reel is the habit of treating the universe as a sequence of stored frames from beginning to present. UED argues that the universe may instead be a structured field of consequences becoming observable through delay, resistance, and alignment.

24. Time is not a cosmic road.

Time is not treated as an independent highway carrying the universe forward. It is causal sequencing, alignment delay, and conscious narration. Events do not happen because time commands them; time appears because consequences unfold through delay.

25. Causality comes before time.

Conditions produce consequences. Time measures and narrates the delay between them. Time does not create causality; causality creates the experience and measurement of time.

26. Alignment Threshold Time explains delay without making time fundamental.

Alignment Threshold Time is the measurable delay required for energy states to rearrange, align, and become observable as consequence. Different field states may carry different alignment delays.

27. Hubble tension may be state-dependent resolution delay.

The disagreement between local and deep-inferred expansion readings may not be cosmic contradiction. It may indicate that different field states resolve through different alignment thresholds.

28. The universe may not be expanding wrongly; we may be timing it too simply.

If redshift carries field strain and time is alignment delay, then apparent expansion rates may include field-resolution effects. Hubble tension may be a diagnostic signal rather than a crisis.

29. The past may be arranged consequence, not a physical archive.

The past is meaningful, but it is known through traces, signals, memory, and interpretation. In cosmology, delayed signals become cosmic history only after the mind and model arrange them.

30. The future is not waiting ahead like a place.

The future is projection built from present understanding and possible consequences. It is real to consciousness as anticipation, but not necessarily an independent region of the universe.

31. The present is where consequence becomes observable.

The present is not merely a clock moment. It is the active interface where a signal, event, or field consequence becomes measurable to observer and instrument.

32. The mind moves through what the field resolves.

The field produces consequences. The mind receives them sequentially, arranges them into time, and experiences motion, history, depth, and change. The universe may be still as totality while the mind experiences flow.

33. Local motion does not prove total cosmic travel.

Planets orbit, galaxies rotate, and signals propagate. Local motion is real. But the universe as totality has no external road, background, or container through which it travels.

34. Change does not require the universe to travel through time.

Change occurs when field conditions transform into consequences. Time measures the delay. The mind narrates the sequence. The universe does not need an independent road to change.

35. Mystery is not the same as confusion.

Mystery belongs to reality. Confusion often belongs to interpretation. The end of cosmic anomalies does not mean the end of mystery; it means the reduction of unnecessary contradiction created by wrong framing.

36. UED does not patch anomalies; it changes the lens.

Universal Energy Dynamics does not add one more exception to the old model. It changes the starting point: field before object, consequence before chronology, alignment before time, saturation before singularity.

37. The universe is not a warehouse of objects.

The cosmos may be better understood as a structured field of energy consequences. Matter, gravity, redshift, time, black holes, and galaxies are connected expressions of deeper field behavior.

38. The vacuum is not absence.

Vacuum is reinterpreted as Structured Vacuum Energy, the active field condition through which matter, gravity, ripple behavior, redshift, saturation, and observable consequences arise.

39. The cosmos does not need repair.

The universe was not broken by JWST, redshift puzzles, black holes, or Hubble tension. The interpretive frame was exposed. When the frame changes, the cosmos becomes coherent again.

40. The end of cosmic anomalies begins when interpretation becomes humble.

The universe is not obligated to fit our intellectual comfort. The model must serve reality, not the other way around. Anomalies invite deeper reading, not defensive patchwork.

Glossary of UED Terms

This glossary explains the core terms used in this book under the framework of Universal Energy Dynamics. These terms are not decorative labels. They are working concepts used to reinterpret matter, gravity, black holes, redshift, time, cosmic depth, and the meaning of anomalies.

The purpose of this glossary is to help the reader hold the UED language clearly before returning to the larger arguments of the book.

Universal Energy Dynamics / UED

Universal Energy Dynamics is the interpretive framework used throughout this book.

UED views existence as a continuous consequence of structured energy, field interaction, ripple behavior, equilibrium-seeking dynamics, alignment, resistance, saturation, and resolution.

It does not begin with dead matter, empty space, or time as an independent cosmic road. It begins with structured energy expressing itself through observable consequences.

Structured Vacuum Energy / SVE

Structured Vacuum Energy is the foundational field-medium in UED.

It is not empty space. It is not absence. It is the structured energetic condition from which matter, gravity, redshift, ripple behavior, saturation, and observable reality arise.

In this view, vacuum is not a passive background. It is the active field condition of manifestation.

Field

A field is the underlying energetic condition within which reality becomes observable.

In UED, the field is not an empty container. It is the active medium through which energy condenses, gradients form, signals propagate, structures stabilize, and consequences become measurable.

The universe is therefore understood as field-first, not object-first.

Field Condition

Field condition refers to the specific state of the field in any given region or relation.

A field condition may be stable, strained, resistant, compact, saturated, aligned, or unresolved. Observed behavior depends not only on objects, but also on the field condition through which those objects and signals emerge.

Field Strain

Field strain refers to tension, pressure, or stress within the field caused by energy differences, resistance, density, compactness, or unresolved gradients.

In this book, redshift is partly reinterpreted as a possible signature of field strain.

Field Resistance

Field resistance is the resistance experienced by ripple behavior, signals, or energy states while resolving through Structured Vacuum Energy.

This resistance may affect redshift, signal delay, luminosity, compactness, and the apparent behavior of cosmic structures.

Field Depth

Field depth means layered difference in field condition.

It includes resistance, strain, energy density, ripple tension, compactness, saturation, and alignment delay. Field depth is not the same as spatial distance alone, and it is not automatically the same as historical past.

A telescope may be resolving field depth, not merely looking backward in time.

Field Realignment

Field realignment is the process through which field conditions reorganize into a new coherent state.

Galaxies, central regulators, redshifted signals, and structured formations may arise through field realignment when energy conditions reach the necessary threshold.

Field Architecture

Field architecture refers to a large-scale organized field structure.

A galaxy, under UED, is not merely a pile of stars, gas, dust, and matter. It is a field architecture expressed through ripple condensation, rotational coherence, luminosity, central regulation, and energy-gradient behavior.

Field Coherence

Field coherence is the condition in which field behavior remains organized enough to produce stable structure, signal, identity, or measurable consequence.

Where coherence is sustained, structure becomes observable. Where coherence fails, signal loss, darkness, or saturation may appear.

Ripple

A ripple is a disturbance, pattern, or propagating behavior within Structured Vacuum Energy.

Ripple behavior is central to UED because matter, signal propagation, field interaction, and saturation are all interpreted through ripple dynamics.

Ripple Condensation

Ripple condensation is the process through which structured energy stabilizes into matter-form.

In UED, matter is not an independent substance. It is condensed ripple behavior within Structured Vacuum Energy.

Ripple Tension

Ripple tension refers to strain or pressure within ripple behavior as it propagates, aligns, resists, or saturates through the field.

Compact cosmic structures, redshifted signals, and saturation zones may carry signs of ripple tension.

Ripple Saturation Zone

A Ripple Saturation Zone is a region where field coherence collapses beyond ordinary observational recovery.

This is UED's reinterpretation of a black-hole-like region. It is not treated as a hole, singularity, or infinite-density object. It is a field condition where ripple information no longer returns in a coherent form to the external observer.

Central Ripple Saturation Zone

A Central Ripple Saturation Zone is a Ripple Saturation Zone located at the core of a galaxy.

It is interpreted as a regulatory saturation condition, not merely as a destructive object. It may help anchor galactic structure, manage field imbalance, and maintain large-scale coherence.

Central Regulator

A central regulator is the dark regulatory core of a galactic field architecture.

What mainstream cosmology often calls a supermassive black hole may, under UED, be understood as a central regulator: an inward saturation boundary that helps hold the outward luminous pattern of the galaxy.

Alignment

Alignment is the process through which energy states or field conditions reach coherence, stability, or observable consequence.

Alignment does not mean emotional harmony or human agreement. It means that a field condition has reached the threshold required for a structure, signal, or consequence to become readable.

Alignment Delay

Alignment delay is the duration required for energy states or field conditions to resolve into observable consequence.

This delay is not empty waiting. It is the active process of resistance, adjustment, saturation, or stabilization before consequence appears.

Alignment Threshold Time / ATT

Alignment Threshold Time is the measurable delay required for energy states to rearrange, align, and become observable as consequence.

ATT is UED's reinterpretation of time. It does not treat time as an independent cosmic road. It treats time as the delay between condition and consequence.

Different field states may have different alignment delays. This is central to the UED interpretation of Hubble tension.

Energy-Gradient Equilibrium

Energy-gradient equilibrium is UED's reinterpretation of gravity.

Where energy density differs, gradients arise. These gradients create directional behavior as the field seeks balance. What we call gravity may be the observable consequence of this imbalance seeking equilibrium.

Gravity, in this sense, does not simply pull. It reveals imbalance.

Consequence

Consequence is the observable outcome of a field condition, energy interaction, or causal process.

In UED, existence itself is understood through consequences. Matter, motion, redshift, gravity, time, and structure are all treated as consequences of deeper field behavior.

Causal Sequencing

Causal sequencing is the order through which conditions lead to consequences.

In UED, causality comes before time. Time does not create causality; causality creates the experience and measurement of time.

State-Dependent Resolution Delay

State-dependent resolution delay means that different field states resolve into observable consequence at different rates.

A stable field, strained field, compact field, or saturated field may not produce the same observable timing. This concept is central to the UED reading of Hubble tension.

Resolution

Resolution is the process by which a field condition becomes observable, measurable, or structurally expressed.

A signal resolves when it becomes detectable. A structure resolves when it becomes coherent. A field state resolves when its consequence becomes readable.

Saturation

Saturation is the condition in which the field reaches a limit beyond which ordinary coherence cannot continue or return in readable form.

A black-hole-like region is reinterpreted as a saturation zone rather than a singular object.

Coherence Boundary

A coherence boundary is the limit beyond which ordinary signal or ripple information can no longer return in recoverable form.

In UED, the event horizon of a black-hole-like region may be better understood as a coherence boundary rather than a magical wall.

Event Horizon

The event horizon is traditionally treated as the boundary beyond which nothing can escape from a black hole.

Under UED, it is reinterpreted as the boundary of recoverable coherence. It marks the point where ordinary ripple information can no longer return in a readable form to the external observer.

Singularity

A singularity is usually described as a point of infinite density where known physical laws break down.

In UED, singularity is not treated as a final physical reality. It is treated as a sign that the model has reached the limit of its interpretive power. UED replaces singularity with saturation.

Matter

Matter is traditionally treated as physical substance.

Under UED, matter is ripple condensation within Structured Vacuum Energy. It is a field consequence, not the first truth of reality.

Mass

Mass is the measurable signature of energy condensation within the field.

In this book, mass is not treated as the ultimate cause of gravity. It is treated as a field-density signature.

Gravity

Gravity is reinterpreted as energy-gradient equilibrium.

It is not merely an invisible pull or the bending of empty spacetime. It is the field's response to imbalance.

Redshift

Redshift is the observed shifting of light toward longer wavelengths.

In mainstream cosmology, redshift is usually interpreted through expansion, distance, and lookback time. UED adds another layer: redshift may also carry field strain, ripple tension, resistance, and alignment delay.

Redshift may not only be a timestamp. It may be the color of resistance.

Cosmic Depth

Cosmic depth refers to the depth of observation into the universe.

In UED, cosmic depth is not reduced only to pastness. It may also mean field depth: layered resistance, strain, compactness, and alignment delay across Structured Vacuum Energy.

Cosmic Maturity

Cosmic maturity is usually interpreted as age-based development.

In UED, maturity may also mean coherence. A galaxy may appear mature because its field architecture is organized and stable, not necessarily because it is old in a conventional chronological sense.

Cosmic Movie Reel

The cosmic movie reel is the metaphor of the universe as a film strip moving from beginning to present.

In this view, distant objects are earlier frames, high redshift is ancient time, and cosmic depth is pastness. UED challenges this metaphor and replaces it with a field-based view of consequences resolving through structured conditions.

Time

Time is not denied in UED. It is repositioned.

Time is real as experience, measurement, duration, and sequence. But it may not be an independent cosmic road. Under UED, time is causal sequencing, alignment delay, and conscious narration.

Past

The past is the mind's arrangement of resolved consequences.

It is meaningful, but it may not exist as an independent cosmic archive. In cosmology, what we call the past is reconstructed from signals, traces, and consequences received in the present.

Future

The future is the mind's projection of possible consequences.

It is real to consciousness as expectation and planning, but it is not necessarily an independent region waiting ahead in the universe.

Present

The present is the active interface where consequence becomes observable.

A signal becomes meaningful when it arrives, is detected, and is interpreted. The present is therefore not only a clock moment. It is the point of resolution.

Signal

A signal is a measurable field consequence reaching observation.

A signal may carry delay, resistance, strain, alignment history, and field-condition imprint. It should not be treated only as a photograph from the past.

Signal Delay

Signal delay is the duration required for a signal to reach observation.

UED does not deny signal delay. It questions whether delay alone should be converted into complete historical certainty.

Hubble Tension

Hubble tension is the disagreement between different methods of estimating the universe's expansion rate.

Under UED, it may be read as evidence of state-dependent resolution delay. Different field states may resolve differently, and those differences may appear as conflicting expansion readings.

Local Motion

Local motion is motion within the universe: planets orbiting, stars moving, galaxies rotating, signals propagating, and structures changing.

UED does not deny local motion. It distinguishes local motion from total cosmic travel.

Total Stillness

Total stillness refers to the idea that the universe as totality may not move through anything outside itself.

Things move within the universe, but the universe as a whole has no external road or background through which it travels.

Conscious Narration

Conscious narration is the mind's arrangement of consequences into past, present, future, identity, memory, expectation, and history.

The field resolves consequences. The mind narrates them as time.

Wholeness

Wholeness means that the universe is not a broken collection of disconnected anomalies.

Under UED, galaxies, black holes, redshift, gravity, time, and Hubble tension are connected expressions of field behavior.

Wholeness does not mean simplicity. It means coherence across layers.

Final Note

The purpose of these terms is not to create a closed doctrine. It is to provide a clearer language for reading cosmic observations beyond the limitations of a matter-first, time-first, movie-reel interpretation.

UED asks the reader to shift the question:

Not merely, "What object is this?"

But, "What field condition is producing this consequence?"

That shift is the beginning of the end of cosmic anomalies.

Quote Bank

- *The universe did not become strange overnight. It only became harder to misread.*
- *An anomaly is often not a failure of reality, but a failure of the frame used to read reality.*
- *JWST did not break the cosmos. It broke the comfort of the cosmic timeline.*
- *The problem is not observation. The problem is interpretation.*
- *A galaxy does not become too early by itself. It becomes too early only inside a timeline that expected immaturity.*
- *Cosmic maturity may mean coherence, not age.*
- *Matter is not a dead thing. It is a field consequence.*
- *Gravity does not pull. Gravity reveals imbalance.*
- *A behaving vacuum cannot be pure nothingness.*
- *Black holes may not be holes. They may be saturation zones.*
- *Darkness is not always absence. Sometimes darkness is saturation.*
- *The event horizon may not be a cosmic wall. It may be the boundary of recoverable coherence.*
- *The galactic center may not be a monster. It may be the regulator of the pattern.*
- *Redshift may not be only the universe's timestamp. It may be the color of resistance.*
- *Looking deeper may not always mean looking backward.*
- *The telescope may not be opening an old archive. It may be resolving field depth.*
- *Time is not the road beneath the universe. Time is the shadow of consequence.*

- *Causality does not need time as ruler. Time needs causality to become meaningful.*
- *The past may be memory of consequence. The future may be projection of consequence. The present is the interface of consequence.*
- *The universe may not be travelling through time. It may be becoming readable through consequence.*
- *The field resolves. The mind narrates.*
- *The cosmic movie reel fails because reality is not obligated to be cinematic.*
- *Hubble tension may not be the universe disagreeing with itself. It may be different field states resolving differently.*
- *The universe may not be expanding wrongly. We may have been timing it too simply.*
- *The end of cosmic anomalies is not the end of mystery. It is the end of unnecessary confusion.*
- *The cosmos was never broken. The frame was.*
- *And when the frame changes, the universe does not need to be repaired. It only needs to be read again.*

Acknowledgments

Every book begins as a question before it becomes a manuscript.

This book began with the question of why the universe is repeatedly being called strange when perhaps our interpretation has become too limited. That question stayed with me through many discussions, reflections, and long internal arguments with the accepted language of cosmology.

I acknowledge the scientific community whose work, observations, instruments, and models have made such questioning possible. Even when this book challenges mainstream interpretation, it does so with respect for the discipline, effort, and intelligence that built the current understanding of the cosmos.

I am grateful to the James Webb Space Telescope mission and to all researchers whose observations have opened deeper questions about galaxies, redshift, black holes, cosmic maturity, and expansion. Their work has strengthened the need to think again.

I also acknowledge my own long journey as a teacher, researcher, and writer. Many ideas in this book have grown through years of studying, teaching, questioning, observing, and refusing to accept easy answers.

My family deserves quiet gratitude for standing with me through this unusual intellectual path. Their patience gives me the space to continue writing and researching.

Finally, I thank the readers who are willing to approach this book with courage and openness. Agreement is not required. Honest reading is enough.

If this book helps even a few minds look at the cosmos with a deeper lens, then its purpose is served.

About the Author

Sandeep J. Chavan is an author, educator, and independent researcher working across physics, philosophy, psychology, human behavior, education, society, and universal theory.

He is the founder of Gyrus Institute of Science, Newasa, where he has spent more than a decade teaching and guiding students in science and competitive examination preparation. His life as a teacher deeply influences his writing style: direct, explanatory, questioning, and accessible to serious readers beyond formal academic boundaries.

As an independent researcher, Sandeep has developed Universal Energy Dynamics, a broad interpretive framework through which he explores energy, matter, gravity, time, consciousness, cosmic structure, and the deeper mechanisms of existence. His work challenges conventional assumptions while trying to build a more integrated understanding of reality through field-based interpretation, consequence, equilibrium, and structured energy behavior.

He has written and published extensively across multiple domains, including scientific reinterpretation, metaphysics, social analysis, self-development, education, and philosophical fiction. His writing often attempts to bridge scientific inquiry with lived human experience, making complex ideas readable without reducing their depth.

The End of Cosmic Anomalies continues his larger intellectual effort to question inherited frameworks and offer a deeper lens through Universal Energy Dynamics.

Disclaimer and Statement of Originality

Disclaimer

This book presents an independent interpretive framework developed by the author under the broader language of Universal Energy Dynamics.

The ideas expressed in this book are intended for intellectual exploration, philosophical inquiry, scientific reflection, and conceptual discussion. They should not be treated as established mainstream scientific consensus, formal institutional doctrine, or a replacement for peer-reviewed cosmological models.

The book discusses topics such as the James Webb Space Telescope, galaxies, black holes, redshift, Hubble tension, time, gravity, matter, and cosmological interpretation. These subjects are actively studied by scientists across the world through observation, mathematical modeling, data analysis, experimentation, and peer-reviewed research. The author acknowledges the value of mainstream science, existing cosmological models, and the contributions of researchers whose work has made modern cosmic inquiry possible.

Where this book questions mainstream interpretation, it does so conceptually and philosophically. It does not claim that every accepted scientific model is useless, nor does it dismiss the discipline of scientific investigation. Rather, it asks whether some observations currently treated as anomalies may require a deeper interpretive lens.

Universal Energy Dynamics, as presented here, is an independent framework. Its terms—including Structured Vacuum Energy, Ripple Saturation Zone, Alignment Threshold Time, field strain, ripple condensation, and central regulator—are used as conceptual tools within the author's system of thought. They should be read as part of this original framework and not confused with officially established terminology in mainstream physics or cosmology unless otherwise stated.

This book does not claim final proof of Universal Energy Dynamics. It presents a structured argument, a philosophical-scientific reinterpretation, and an invitation for deeper inquiry. Readers are encouraged to think critically, compare ideas, consult mainstream scientific sources, and approach the subject with both openness and discipline.

The purpose of this book is not to replace scientific method with belief. It is to encourage questioning where interpretation appears too narrow for observation.

Statement of Originality

The conceptual framework, language, arguments, interpretations, and narrative structure presented in this book are original to the author's work within the Chavanian Ecosystem and Universal Energy Dynamics.

The book develops an independent reinterpretation of cosmic anomalies by connecting galaxies, black holes, redshift, Hubble tension, time, matter, gravity, and cosmic maturity through a field-based, consequence-based lens. Its central claim is that many so-called anomalies may not be failures of the universe, but failures of interpretation.

Key ideas presented in this work include:

Matter as ripple condensation within Structured Vacuum Energy.

Gravity as energy-gradient equilibrium.

Black-hole-like regions as Ripple Saturation Zones.

Galactic centers as central regulators.

Redshift as field strain, ripple tension, resistance, and alignment delay.

Time as Alignment Threshold Time and conscious narration.

Hubble tension as state-dependent resolution delay.

Cosmic depth as field depth, not merely pastness.

Cosmic maturity as coherence, not only chronological age.

These ideas are developed as part of the author's original intellectual contribution. They are not presented as copied extensions of any single existing school of physics, cosmology, metaphysics, or philosophy. While the book engages with publicly known scientific concepts and observations, the interpretive synthesis belongs to the author's independent framework.

Any resemblance between specific terms, phrases, or conclusions and other works is either coincidental or arises from shared engagement with broad scientific and philosophical questions. The author has made a conscious effort to present the ideas in his own language, structure, and conceptual system.

This work is written with respect for science, but also with the freedom of independent inquiry. It stands as an original contribution to the ongoing human effort to understand reality, not as a final declaration, but as a serious invitation to reconsider the frame through which the universe is read.

The cosmos does not need our permission to be what it is.

But our interpretations must remain humble enough to change.

Further Reading

From the Chavanian Ecosystem

The End of Cosmic Anomalies belongs to the larger Chavanian Ecosystem, a body of work that explores reality through structure, consequence, energy, awareness, field behavior, systems, and human experience.

This book focuses on cosmology, but its foundation is connected with several other works that develop the same language from different directions. Readers who want to go deeper into Universal Energy Dynamics, redshift, light, time, uncertainty, persistence, manifestation, and field-based reality may begin with the following books.

Universal Energy Dynamics

This is the foundational work behind the framework used throughout *The End of Cosmic Anomalies.*

Universal Energy Dynamics introduces a consequence-based understanding of reality through constraints, gradients, thresholds, imbalance, and resolution. It provides the broader foundation for the ideas developed in this book: matter as field consequence, gravity as energy-gradient behavior, time as alignment delay, and reality as structured energy in motion.

Readers who want to understand the complete UED framework should begin here.

Redshift: Beyond Expansion

This is the most directly connected companion to the present book.

Redshift: Beyond Expansion re-examines one of cosmology's most important observations. It does not reject redshift, but questions whether redshift should be interpreted only through expansion, distance, and lookback time.

Readers interested in Chapter 8 of this book should continue with this title. It expands the idea that redshift may carry deeper meanings related to observation, resolution, field strain, and the hidden geometry of the universe.

Photon Reframed

This book is especially relevant for readers interested in light, signal, and observation.

Photon Reframed questions whether the photon should be understood as a fixed particle or wave. It offers a process-based interpretation of light, energy, and reality, and connects naturally with this book's treatment of redshift, signal delay, ripple tension, and field resolution.

Readers who want to explore the deeper nature of light after reading about redshift and JWST should read this next.

Delta – The Structure of Change

Delta explores change through structure, imbalance, and resolution.

It is closely connected to the UED view that reality does not simply move through time but resolves through field conditions. The book is useful for understanding why change should not be treated as random motion or simple chronological development, but as a consequence of structured imbalance seeking resolution.

Readers interested in Chapters 9, 10, and 11 of this book will find *Delta* especially useful.

The Law of Persistence

This book studies the hidden structure behind what remains stable, continuous, and enduring.

In relation to *The End of Cosmic Anomalies*, it helps readers understand why persistence, coherence, and stability matter more deeply than surface motion. Galaxies, field structures, systems, and human realities persist not merely because time passes, but because certain structural conditions remain active.

Readers interested in cosmic coherence and field stability should continue with this title.

Uncertainty – The Pre-Resolved Strain

This work reframes uncertainty as a structural condition rather than mere ignorance, randomness, or error.

It connects strongly with this book's treatment of Hubble tension, redshift uncertainty, observational limits, and field-state differences. Uncertainty is not always a weakness in knowledge; it may be the signature of systems operating under dynamic equilibrium before full resolution.

Readers interested in Hubble tension and Alignment Threshold Time should read this as a companion work.

Manifestation: From Nothing to Everything

This book explores existence, reality, and the process through which the unmanifest becomes observable.

It is useful for readers who want to understand the deeper philosophical foundation behind UED. While *The End of Cosmic Anomalies* focuses on cosmological interpretation, *Manifestation* expands the larger question: why does anything exist, and how does reality become structured?

Readers who want the metaphysical foundation behind field, energy, and consequence should continue here.

Life: The Sustained Imbalance

This short manifesto presents life as sustained imbalance within constraint.

Though focused on life, it carries the same structural language used in UED: imbalance, constraint, resolution, and continuity. It helps readers see that the same logic used to interpret the cosmos can also apply to living systems.

Readers who want a concise bridge between cosmic structure and living reality may read this work.

Soul Reframed

Soul Reframed explores life, mind, and identity through the field of Structured Vacuum Energy.

It connects the cosmic side of UED with the human side. After reading about matter, energy, and field behavior in the universe, readers may find this book useful for understanding how similar principles can be applied to life, identity, and consciousness.

Awareness: What Cannot Be Engineered

This book explores the difference between intelligence and awareness in an age shaped by AI, systems, optimization, and distraction.

While it is not a cosmology book, it connects with the later chapters of *The End of Cosmic Anomalies*, especially where the book discusses conscious narration, time, past, future, and the moving mind.

Readers interested in the observer's role in interpretation should continue with this title.

When Intelligence Outruns Awareness

This book studies the gap between capability and clarity.

It is relevant because many scientific and technological systems become powerful before interpretation matures. This is exactly the situation presented in *The End of Cosmic Anomalies*: instruments have become sharper, but interpretation may still be trapped inside older assumptions.

Readers interested in why intelligence can create overconfidence should read this book.

Suggested Reading Path

Readers who want to continue directly from *The End of Cosmic Anomalies* may follow this path:

1. *Redshift: Beyond Expansion* For a deeper study of redshift, observation, and interpretation.
2. *Photon Reframed* For a process-based understanding of light, signal, and energy.
3. *Universal Energy Dynamics* For the complete framework behind consequence-based reality.
4. *Delta – The Structure of Change* For change, imbalance, motion, and resolution.
5. *Uncertainty – The Pre-Resolved Strain* For uncertainty, tension, and unresolved field states.
6. *The Law of Persistence* For stability, continuity, and the structure behind what endures.
7. *Manifestation: From Nothing to Everything* For the larger philosophical foundation of existence and reality.

Closing Note

The Chavanian Ecosystem is not a collection of isolated books. It is an expanding intellectual architecture.

Some books begin from the cosmos.

Some begin from life.

Some begin from mind.

Some begin from society, awareness, systems, or human behavior.

But beneath all of them is one recurring question:

What condition produces the consequence we are observing?

The End of Cosmic Anomalies asks this question at the scale of the universe. The books listed here continue the same inquiry through light, redshift, time, uncertainty, persistence, manifestation, life, and awareness.

Together, they invite the reader to move from surface explanation to structural understanding.

Don't miss out!

Visit the website below and you can sign up to receive emails whenever Sandeep Chavan publishes a new book. There's no charge and no obligation.

https://books2read.com/r/B-A-EPGPC-CQZJJ

BOOKS 2 READ

Connecting independent readers to independent writers.

Did you love *The End of Cosmic Anomalies*? Then you should read *Universal Energy Dynamics*[1] by Sandeep Chavan!

[2]

Universal Energy Dynamics: A Consequence-Based Framework for Reality

Why do systems fail even when intelligence, technology, and planning are present?Why do crises appear suddenly after long periods of apparent stability?Why do actions produce consequences far removed from their origin?

In *Universal Energy Dynamics*, independent researcher and author **Sandeep J. Chavan** introduces a powerful interpretive framework for understanding how complex systems behave across physics, human institutions, and global networks.

Rather than explaining events through isolated causes or intentions, the book proposes a deeper structural model based on consequence-driven dynamics. According to this framework, systems evolve through the

1. https://books2read.com/u/mdBOnw

2. https://books2read.com/u/mdBOnw

interaction of constraints, accumulated pressure, gradients of change, and threshold-driven correction.

At the center of Universal Energy Dynamics (UED) are several foundational concepts:

• **Delta** — unresolved pressure accumulating within a system• **Gradient** — directional differences that drive resolution• **Constraint** — limits that shape how systems evolve• **ATT (Alignment Threshold Time)** — the tolerance window before abrupt correction• **Consequence** — the structural feedback that restores equilibrium

Together, these principles provide a diagnostic lens for analyzing events that often appear chaotic or unpredictable.

Through conceptual explanations, diagrams, and real-world case studies—including financial crises, organizational failures, pandemics, climate systems, and information networks—**Sandeep J. Chavan** demonstrates how the same underlying dynamics operate across vastly different domains.

The book does not reject existing scientific theories. Instead, it places them within a broader structural perspective, showing how classical physics, relativity, quantum mechanics, and thermodynamics describe outcomes while Universal Energy Dynamics explores the deeper mechanisms that drive them.

By shifting focus from intention to consequence, this framework reveals why:

• stability can hide accumulating instability• systems fail after long periods of success• tightly connected networks amplify pressure• delayed feedback creates the illusion of randomness

Written in clear, disciplined language, *Universal Energy Dynamics* bridges the gap between science, philosophy, and systems thinking. It offers readers a new way to interpret events—from organizational decisions to planetary-scale challenges—through the lens of structural dynamics.

This book is not a manifesto or a prediction engine. It is a tool for understanding how systems accumulate pressure, cross thresholds, and correct themselves through consequence.

For readers interested in **systems science, philosophy of physics, complexity theory, and interdisciplinary thinking**, *Universal Energy Dynamics* provides a compelling framework for seeing reality in a new way.

Read more at https://www.sandeepjchavan.com/.

www.ingramcontent.com/pod-product-compliance
Lightning Source LLC
LaVergne TN
LVHW100723061126
851857LV00011B/373

* 9 7 8 8 1 6 8 9 7 7 6 8 6 *